Evelin Wacker

Tactile Feature Processing and Selective Attention

Evelin Wacker

Tactile Feature Processing and Selective Attention

Two Experimental Investigations in the Human Somatosensory System

Südwestdeutscher Verlag für Hochschulschriften

Impressum/Imprint (nur für Deutschland/only for Germany)
Bibliografische Information der Deutschen Nationalbibliothek: Die Deutsche Nationalbibliothek verzeichnet diese Publikation in der Deutschen Nationalbibliografie; detaillierte bibliografische Daten sind im Internet über http://dnb.d-nb.de abrufbar.
Alle in diesem Buch genannten Marken und Produktnamen unterliegen warenzeichen-, marken- oder patentrechtlichem Schutz bzw. sind Warenzeichen oder eingetragene Warenzeichen der jeweiligen Inhaber. Die Wiedergabe von Marken, Produktnamen, Gebrauchsnamen, Handelsnamen, Warenbezeichnungen u.s.w. in diesem Werk berechtigt auch ohne besondere Kennzeichnung nicht zu der Annahme, dass solche Namen im Sinne der Warenzeichen- und Markenschutzgesetzgebung als frei zu betrachten wären und daher von jedermann benutzt werden dürften.

Coverbild: www.ingimage.com

Verlag: Südwestdeutscher Verlag für Hochschulschriften GmbH & Co. KG
Heinrich-Böcking-Str. 6-8, 66121 Saarbrücken, Deutschland
Telefon +49 681 37 20 271-1, Telefax +49 681 37 20 271-0
Email: info@svh-verlag.de

Approved by: Berlin, TU, Dissertation, 2011

Herstellung in Deutschland:
Schaltungsdienst Lange o.H.G., Berlin
Books on Demand GmbH, Norderstedt
Reha GmbH, Saarbrücken
Amazon Distribution GmbH, Leipzig
ISBN: 978-3-8381-1617-4

Imprint (only for USA, GB)
Bibliographic information published by the Deutsche Nationalbibliothek: The Deutsche Nationalbibliothek lists this publication in the Deutsche Nationalbibliografie; detailed bibliographic data are available in the Internet at http://dnb.d-nb.de.
Any brand names and product names mentioned in this book are subject to trademark, brand or patent protection and are trademarks or registered trademarks of their respective holders. The use of brand names, product names, common names, trade names, product descriptions etc. even without a particular marking in this works is in no way to be construed to mean that such names may be regarded as unrestricted in respect of trademark and brand protection legislation and could thus be used by anyone.

Cover image: www.ingimage.com

Publisher: Südwestdeutscher Verlag für Hochschulschriften GmbH & Co. KG
Heinrich-Böcking-Str. 6-8, 66121 Saarbrücken, Germany
Phone +49 681 37 20 271-1, Fax +49 681 37 20 271-0
Email: info@svh-verlag.de

Printed in the U.S.A.
Printed in the U.K. by (see last page)
ISBN: 978-3-8381-1617-4

Copyright © 2011 by the author and Südwestdeutscher Verlag für Hochschulschriften GmbH & Co. KG and licensors
All rights reserved. Saarbrücken 2011

Acknowledgements

The research for this doctoral thesis was carried out in the Junior Research Group "Neuroimaging and Neurocomputation" headed by Dr. Felix Blankenburg. Besides him, many people supported me professionally as well as personally throughout my research. Without their help, this thesis would not have been accomplished.

First, I would like to thank Dr. Felix Blankenburg for providing me the opportunity to do my doctoral research in his group and for supporting me in all aspects of my work. I very much appreciated his excellent supervision and his inspiring ideas, on which I gratefully seized for my research. Many thanks further go to Prof. Dr. Felix Wichmann, who provided me helpful advice and supervised my work as second expert.

I also want to thank Prof. Dr. John-Dylan Haynes and Dr. Jakob Heinzle, who composed, along with Dr. Felix Blankenburg and Prof. Dr. Felix Wichmann, my PhD Committee, which offered me valuable support in the course of my research project. Further, I am very thankful to Dr. Vanessa Casagrande, Julia Schaeffer, and Margret Franke for their patience with all the questions concerning the research training group and organizational matters. Additionally, I gratefully acknowledge funding from the BMBF and the GRK 1589/1 "Sensory Computation in Neural Systems".

Special thanks go to my team mate and collaborator Dr. Bernhard Spitzer, who always encouraged me and taught me so much. I am also very thankful to my colleagues Ryszard Auksztulewicz, Dr. Dirk Ostwald, and Bianca van Kemenade for fruitful discussions and daily enjoyments during the time in our lab. Moreover, I would like to express my thanks to my collaborators Prof. Dr. Johannes Bernarding and Ralf Lützkendorf, who provided me the opportunity and technical support to use the 7 Tesla MRI in Magdeburg.

I would further like to thank Dr. Bernhard Spitzer, Uwe Benary, and Dr. Juliane Klein for critically revising this manuscript. Their suggestions contributed to its improvement substantially. Finally, I am very grateful to my friends, in particular Juliane and Manuela, for their patience and their constant encouragement. In addition, I won't forget to thank my parents for creating the firm foundation I stand on and for giving me all the love they have. And last, but definitely not least, I would like to express my deepest gratitude to Christian for his understanding, for all his love, and for his constant faith in me.

Abstract

The somatosensory system offers us diverse functionality. It is responsible for the sensation of touch, which involves perception of external objects. It provides information about the body's own components and is critically involved in the planning and execution of motor actions. Considerable effort has been directed towards elucidating how somatosensory processing is organized to subserve these various functions. Based on this, Dijkerman and de Haan proposed a model to describe the cortical processing of somatosensory information (Dijkerman and de Haan 2007). The aim of the present thesis was to extend this model of somatosensory processing for perception and action, focussing on feature processing and attentional modulation during tactile perception. To this end, two functional magnetic resonance imaging experiments were performed, in which Braille-like tactile stimulation was presented to human volunteers. The first experiment sought to determine the role of feature-specific higher-order processing for tactile perception and involved moving or patterned stimulation during passive touch. We found that the visual motion-sensitive area hMT+/V5 and the inferior parietal cortex were selectively activated during motion and pattern processing, respectively. The responses covaried with participants' perceptual performance in identifying the respective stimulus attribute and were functionally coupled to the responses in primary somatosensory cortex. The results of this study provided evidence for the direct engagement of feature-specific cortical areas in tactile perception. The second experiment aimed at investigating the functional significance of top-down attentional gating during tactile task accomplishment. The task involved selective attention to the spatial pattern or to the temporal frequency of the tactile stimulation and the detection of changes in the respective stimulus attribute. We found that a frontoparietal network was selectively activated during the detection of task-relevant change. Analysis of effective connectivity revealed that the functional integration of task-relevant sensory information occured in a network composed of the somatosensory cortices and the inferior frontal gyrus. Modeling context-dependent causal influences within this functional network identified top-down attentional biasing for gating perception of tactile stimulus attributes. Based on the findings presented here, functions for feature processing and attentional modulation were added to the model by Dijkerman and de Haan. The extended model contributes to the understanding of how the somatosensory system processes tactile input and allows formulating testable hypotheses to motivate future research questions.

Parts of the present thesis have been published as peer-reviewed journal article: Wacker E, Spitzer B, Lützkendorf R, Bernarding J, and Blankenburg F (2011). Tactile motion and pattern processing assessed with high-field fMRI. *PLoS One*, 6:e24860.

Contents

1. Introduction   7

2. The Human Somatosensory System   11
   - 2.1. Peripheral Receptors for Tactile Perception 11
   - 2.2. Cortical Regions Involved in Somatosensory Processing 13
   - 2.3. Tactile Features and Feature Representation 16
   - 2.4. The Model of Somatosensory Processing for Perception and Action 17
   - 2.5. Attention in the Somatosensory System 18

3. Introduction to Functional Magnetic Resonance Imaging   23
   - 3.1. Physics of Magnetic Resonance Imaging 23
   - 3.2. Hemodynamics and Magnetic Resonance 24
   - 3.3. Statistical Analysis of Functional Imaging Data 25
      - 3.3.1. Image Preprocessing 25
      - 3.3.2. Statistical Analysis 26
   - 3.4. Functional Integration and Effective Connectivity 28
      - 3.4.1. Psychophysiological Interaction Models 29
      - 3.4.2. Dynamic Causal Modeling 29

4. Feature-Specific Processing of Tactile Stimulus Attributes   35
   - 4.1. Introduction 35
   - 4.2. Materials and Methods 38
   - 4.3. Results 44
   - 4.4. Discussion 49

5. Top-Down Attentional Bias for Gating Tactile Perception   55
   - 5.1. Introduction 55
   - 5.2. Materials and Methods 58

Contents

    5.3. Results 63
    5.4. Discussion 70

6. Summary and Model Extension   **77**
    6.1. Summary of Experimental Results 77
    6.2. The Extended Model of Somatosensory Processing 78
    6.3. Outlook 82

7. Conclusions   **85**

A. BOLD Time Courses   **87**

B. Additional Figures   **89**

C. Bootstrapping Correlations   **91**

Bibliography   **95**

List of Figures   **109**

List of Tables   **111**

List of Abbreviations   **113**

Remark on Word Usage
In accordance with the standard scientific protocol, the personal pronoun "we" will be used to indicate the reader and the writer, or the author and her scientific collaborators.

Chapter 1.

Introduction

Touch as one of the traditional five senses informs the organism - along with sight, hearing, smell, and taste - about the nature of things in the external world. Be it a fruit's maturity, the quality of sheer fabric, or the softness of baby's skin - all these are conscious perceptual experiences of touch. Touch is one of the sensory entities that make up the somatosensory system, which by itself is highly diverse in its functionality. Touch supplies the organism with inputs from the external palpable world, whereas proprioception provides information about the internal state of the body and the relative position of body parts. Both functions of the somatosensory system are closely tied and act in concert to create the prerequisites for perception and action.

How the brain subserves somatosensory processing for tactile perception is still subject to research. Tactile input from the peripheral receptors covering the skin passes via sensory nerve fibers through the dorsal column in the spinal cord to the brain. The existence of several somatotopically organized cortical areas of somatosensory representations was revealed by electrophysiological studies as early as in the 1940s (Penfield and Boldrey 1937; Woolsey 1943). Brodmann areas 3a, 3b, 1, and 2 of the postcentral gyrus constitute the primary somatosensory cortex (SI). The secondary somatosensory cortex (SII) is located in the parietal operculum in the upper bank of the lateral fissure and by itself comprises several subregions. In analogy to the processing hierarchy in the visual system it was proposed that sensory information is propagated from the thalamus to SI and then to SII. In fact, SI seems to be mainly engaged in early processing of relatively simple features such as the location or the duration of a tactile stimulus. Subsequent processing may involve more complex features such as the direction or the velocity of an object moving along the skin. Higher-order association areas may combine these features to provide information about the shape of an object or integrate it in a representation

Chapter 1. Introduction

of the body (Dijkerman and de Haan 2007). Based on existing ideas about the organization of the cortical somatosensory system, Dijkerman and de Haan recently proposed a model of somatosensory processing for perception and action. According to this model, somatosensory processes for the guidance of action can be dissociated from those that lead to perception and memory (Dijkerman and de Haan 2007). This segregation can be considered analogous to the well-established "what" versus "where" distinction of information processing in the visual system along the ventral and dorsal streams (Mishkin and Ungerleider 1982). Nevertheless, how cortical somatosensory processing is organized in detail to subserve tactile perception is far from understood compared with the visual system.

The aim of the present thesis is to extend the model of somatosensory processing for perception and action proposed by Dijkerman and de Haan, focussing on conscious perception of complex tactile stimuli and their properties. Of particular interest is the impact of feature-specific higher-order processing on tactile perception but also the functional significance of top-down attentional gating during tactile task accomplishment. Both higher-order and top-down processing became more and more important during the last years. As regards the former, there is increasing evidence that feature-specific processing of complex sensory input may generalize across modalities and result in a higher-order perceptual representation of abstract features. As an example, the lateral occipital complex (LOC) appears to hold a multisensory representation of shape during visual and tactile object recognition (e.g., Amedi et al. 2001; Lacey and Sathian 2011). In terms of top-down processing, the increasing importance of top-down influences on sensory perception is reflected in the development of theories such as predictive coding (Rao and Ballard 1999). According to the predictive coding theory, prior information such as expectation, memory, or knowledge may exert top-down influence on sensory areas during perceptual inference to facilitate the interpretation of sensory input. Here, we aim at investigating feature processing and attentional top-down modulation during tactile perception and further at incorporating these principles into the model of somatosensory processing for perception and action.

For this purpose, we performed two functional magnetic resonance imaging (fMRI) experiments with healthy human volunteers. Tactile stimuli with different properties were applied to the index finger tip of the participants' hand and associated with different cognitive task-related requirements. The suggested stimuli were created using a programmable stimulation device (Piezostimulator, QuaeroSys) equipped with a 4x4 Braille-like tactile display that allowed administering well-described and replicable cutaneous input within a circumscribed area of glabrous skin.

The first experiment focussed on investigating the impact of feature-specific higher-order processing on tactile perception. For this purpose, the tactile stimuli were applied under passive touch and involved different features, namely motion and pattern. The passive stimulation paradigm allowed us to look into bottom-up sensory processing during functional integration in the somatosensory network. We further assessed the effective connectivity between feature-specific cortical areas and somatosensory cortex using psychophysiological interaction (PPI) analyses (Friston et al. 1997). The results of this study form the first extension of the model of somatosensory processing.

In the second experiment, similar tactile stimuli were used but participants' cognitive context was varied during tactile stimulation. A tactile change-detection task that involved selective attention to a specific tactile stimulus attribute, in this case pattern or frequency, required both the deployment of top-down directed attention and the identification of task-relevant bottom-up sensory information. Based on the experimental data, we established network models using dynamic causal modeling (DCM; Friston et al. 2003) to reveal the causal relationships between prefrontal cortex and sensory areas during task performance, and to examine the coordination of bottom-up sensory processing and top-down attentional control. The results of this study constitute the second extension of the model of somatosensory processing.

The findings resulting from the two experimental investigations are incorporated into the existing model of somatosensory processing for perception and action. This extended model complements the current knowledge about somatosensory processing for conscious perception and contributes to the understanding of the human somatosensory system.

Structure of the Present Thesis and Author's Contributions

The theoretical basics that help to approach the topic are reviewed in the two chapters following this introduction. Chapter 2 gives an overview of the essential neurobiological background introducing the human somatosensory system. Peripheral receptors and cortical regions devoted to the somatosensory system are described as well as general features of tactile stimuli and their representation in the somatosensory system. The model of somatosensory processing for perception and action proposed by Dijkerman and de Haan is presented in detail. This model motivated the experimental investigations presented in this thesis. An additional paragraph is dedicated to the role of attention in the somatosensory network. Chapter 3 introduces the technique of fMRI, and describes the

Chapter 1. Introduction

methods for statistical data analysis used in the present thesis. These include the general linear model for conventional statistical inference as well as PPI models and DCM for the analysis of functional integration.

The experimental investigations are presented in detail in Chapters 4 and 5. Both chapters start by introducing the current knowledge in the field and motivate the respective research questions. A detailed description of the applied methods follows as well as a comprehensive presentation of the results. Conclusions are drawn subsequently and their significance in the field is evaluated. Chapter 4 deals with feature-specific higher-order processing of tactile stimulus attributes, Chapter 5 focusses on top-down attentional gating of somatosensory perception. Both experiments were conceptualized and designed by the author including the tactile stimuli used. The author's contributions further comprise data acquisition and analysis as well as the interpretation of the results. The findings provide new insights into somatosensory processing and complement the knowledge about the human somatosensory system.

In Chapter 6 the findings resulting from the two fMRI experiments are summarized and incorporated into the model of somatosensory processing for perception and action. This extended model provides a new description of how the cortical somatosensory system is organized to subserve tactile perception and object recognition. We further give an outlook on future research questions that may build on the proposed model. Chapter 7 completes the thesis by drawing conclusions.

Chapter 2.

The Human Somatosensory System

A major function of the nervous system is to gather information about the external world. This task is performed by different sensory systems where sight, hearing, touch, smell, and taste originate that form the basis for the knowledge about the world. Sensory input is transformed by specialized receptor cells into an electrical signal. These receptor cells are linked along different afferent nerve fibers to the cerebral cortex where their input is processed. The somatosensory system is responsible for the sensation of touch.

This chapter gives an overview of the human somatosensory system. We start by introducing its organization focussing on peripheral receptors for tactile perception and cortical regions dedicated to somatosensory processing. In addition, we consider general features of tactile stimuli as well as their representation in the somatosensory system and introduce the model of somatosensory processing for perception and action. A further section is dedicated to the role of attention in the somatosensory system.

The interested reader is referred to Kandel et al. 2000 for general background information on neurobiological principles. For more details on the somatosensory system, we recommend Mountcastle 2005 and Nelson 2002, or refer to the references indicated in this chapter.

2.1. Peripheral Receptors for Tactile Perception

In the somatosensory system, incoming information about the external world is subdivided into separate processing streams. This division already begins at the level of peripheral receptors. Four types of somatosensory receptors, namely cutaneous mechanoreceptors, proprioceptors, thermoreceptors, and nociceptors, are involved in the transmission of different stimulus properties. Cutaneous mechanoreceptors are responsible for tactile per-

Chapter 2. The Human Somatosensory System

Afferent	Receptor	RF size	Adaptation rate	Function
SAI	Merkel	Small	Slow	Form, texture
SAII	Ruffini	Large	Slow	Lateral force, movement
RAI	Meissner	Small	Rapid	Velocity, flutter (best at 20-30 Hz)
RAII	Pacinian	Large	Rapid	Movement, vibration (best at 250 Hz)

Table 2.1.: Overview of mechanoreceptive afferents and their properties. SAI, slowly adapting type I; SAII, slowly adapting type II; RAI, rapidly adapting type I; RAII, rapidly adapting type II; RF, receptive field.

ception, whereas proprioceptors, thermoreceptors, and nociceptors transduce information about the internal state of the body, temperature, and potentially damaging stimuli, respectively. The present thesis focusses on tactile perception, which relies on different types of mechanoreceptive afferents innervating the glabrous skin. Their neural response properties have been studied extensively in both human and nonhuman primates. In the following, we shortly present the four types of cutaneous mechanoreceptors that each have distinct functions and in concert account for tactile perception (for review see Johnson et al. 2000). An overview is given in Table 2.1.

Slowly Adapting Mechanoreceptive Afferents
Two of the four cutaneous mechanoreceptive afferent types comprise slowly adapting type I and II (SAI and SAII) afferents, which are connected to Merkel cells and Ruffini endings, respectively. They were classified as slowly adapting because they respond to sustained skin deformation with a sustained discharge that declines slowly (Johnson 2001). SAI afferents have small receptive fields leading to high-resolution spatial neural images of tactile stimuli. An important property is surround suppression (Vega-Bermudez and Johnson 1999), which results in strong responses to points, edges, and curvature but weak responses to uniform skin indentation. Therefore, local spatial features such as edges and curves are represented by the peripheral SAI population response. Neurophysiological evidence further indicates that the SAI system plays an important role in form and texture perception (Johnson and Hsiao 1992).

SAII afferents have larger receptive fields than SAI afferents. They are less sensitive to cutaneous indentation but more sensitive to skin stretch (Johnson 2001). Therefore,

SAII afferents are suited to signal lateral forces such as pulling on an object held in the hand. Their sensitivity to skin stretch is further important for the perception of hand conformation and of the direction of an object moving across the skin (Johnson et al. 2000).

Rapidly Adapting Mechanoreceptive Afferents

The remaining cutaneous mechanoreceptive afferent types are rapidly adapting type I and II (RAI and RAII) afferents, which are connected to Meissner and Pacinian corpuscles, respectively. They were classified as rapidly adapting because they respond only transiently to sudden, steady indentation (Johnson 2001). Compared to SAI afferents, RAI afferents innervate the skin of the finger pad more densely, and their receptive field sizes are larger, resulting in great sensitivity but poor spatial resolution. The SAI and RAI response properties are complementary. An important characteristic of RAI afferents is their sensitivity to motion across the skin. Neurophysiological studies show that the RAI system plays a significant role in the perception of low-frequency motion (for review see Johnson et al. 2000), including the detection of low-frequency vibration and slip.

RAII afferents are extremely sensitive but the boundaries of their receptive fields are difficult to determine (Bell et al. 1994). The receptive field of a single RAII afferent may be of any size; it may be restricted to a single phalanx, or encompass the entire hand or arm, depending on its sensitivity. Due to the small number of RAII afferents and their large receptive fields, the RAII system represents little information about the spatial properties of a tactile stimulus. Instead, it is responsible for the perception of high-frequency vibrations transmitted through objects contacting the skin (Johnson et al. 2000).

2.2. Cortical Regions Involved in Somatosensory Processing

Sensory information from the cutaneous mechanoreceptors is transmitted via the dorsal column in the spinal cord to the medulla oblongata, i.e., the lower part of the brainstem. The input crosses to the other side of the brain and is then transferred to the ventroposterior nucleus (VP) of the contralateral thalamus. Somatosensory cortical areas that receive direct thalamo-cortical projections are the primary somatosensory cortex (SI), and, to a lesser extent, the secondary somatosensory cortex (SII).

Chapter 2. The Human Somatosensory System

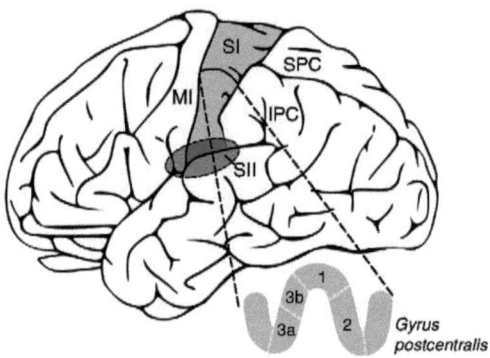

Figure 2.1.: Cortical regions involved in somatosensory processing. SI is located in the postcentral gyrus of the anterior parietal cortex comprising areas 3a, 3b, 1, and 2 (see enlarged cutout). SII can be found in the upper bank of the lateral sulcus. Flanking cortical regions are labeled. MI, primary motor cortex; SPC, superior parietal cortex; IPC, inferior parietal cortex.

Primary Somatosensory Cortex

SI is located in the postcentral gyrus of the anterior parietal cortex and comprises four cytoarchitectonically distinct subregions arranged from anterior to posterior (Figure 2.1). These regions are termed areas 3a, 3b, 1, and 2, according to the classification of Brodmann (Brodmann 1909). SI contains several somatotopic maps of the contralateral half of the body referred to as "sensory homunculus", which were first identified in the human brain (Penfield and Boldrey 1937) but also in several nonhuman primates (Merzenich et al. 1978; Kaas et al. 1979; Nelson et al. 1980). Thalamo-cortical projections end predominantly in areas 3b and 3a, wheras areas 1 and 2 receive much less direct input from the thalamus. Instead, areas 1 and 2 receive direct projections from areas 3b and 3a. This hierarchy in sensory information processing resulted in the view that only area 3b can be considered to be the homologue of SI in non-primates, which is referred to as "SI proper" (Kaas 1983). There is further evidence for a functional specialization within SI, according to which neurons in areas 3a and 2 respond primarily to proprioceptive stimulation, and neurons in areas 3b and 1 are more sensitive to mechanical stimulation (Powell and Mountcastle 1959). However, this functional segregation is not mutually exclusive as areas 1 and 2 integrate both tactile and proprioceptive information.

Secondary Somatosensory Cortex

SI is reciprocally connected to SII, which is located in the parietal operculum in the upper bank of the lateral sulcus (Figure 2.1). The bilateral receptive fields of SII are larger compared to those of SI. Somatotopic representations of the ipsi- and contralateral body half were found in several subdivisions of SII but they are less detailed than those in SI. The parietal ventral area (PV), the ventral somatosensory area (VS), and area S2 are well preserved subregions of SII that can be found in many nonhuman species (Cusick et al. 1989; Krubitzer and Kaas 1990; Krubitzer et al. 1995). With respect to the human brain, early stimulation studies were limited in their description of SII due to the deep location in the lateral sulcus and the technique's insufficient resolution to describe the organization of SII in any detail. Initial evidence for several somatotopically organized areas within the human SII came from a functional imaging study (Disbrow et al. 2000). Postmortem investigations of the human parietal operculum revealed the existence of four distinct cytoarchitectonic subregions (Eickhoff et al. 2006a,b), which were termed OP 1 to OP 4 (OP stands for *operculum parietale*). OP 4, 1, and 3 constitute the putative human homologues of areas PV, S2, and VS, respectively.

Other Somatosensory Areas

SII is reciprocally connected to granular and dysgranular fields of the insula. Friedman and colleagues reasoned that this connection constitutes a cascaded pathway to brain structures essential for tactile learning and memory (Friedman et al. 1986). Neurophysiological recordings in monkeys further showed that a major portion of this area is involved in somatosensory information processing (Schneider et al. 1993).

Furthermore, the posterior parietal cortex (PPC) is related to somatosensory processing. Several foci in the PPC receive direct projections from both SI and SII but also from the thalamus. The PPC can be subdivided into inferior parietal cortex (IPC) and superior parietal cortex (SPC), which are separated by the intraparietal sulcus (IPS). Anterior parts of the PPC are implicated in the spatio-temporal integration of tactile input as well as in somato-motor control (e.g., Binkofski et al. 1999; Bodegård et al. 2001; Van Boven et al. 2005). The PPC further receives input from different sensory modalities and appears to play a key role in multisensory integration for higher-level processing (Bremmer et al. 2001). The detailed functional organization of the PPC remains to be elucidated.

An interesting question is whether a functional dissociation of two processing streams similar to the two-visual-systems hypothesis (Mishkin and Ungerleider 1982) holds also true in the somatosensory system. It was suggested that the processing stream from

SI via SII to the insula involved in tactile perception and learning may represent the analogue of the ventral "what" pathway in the visual system (Mishkin 1979; Friedman et al. 1986). Projections from SI, either directly or via SII, to the PPC are implicated in action-related processing. Therefore, this pathway was suggested to constitute the tactile dorsal equivalent involved in the processing of information about "where" or "how" (Reed et al. 2005). This proposition is elaborated on in Section 2.4. The dissociation of object and spatial processing streams seems to be a general principle of sensory systems, which is reflected in functional specialization in the brain.

2.3. Tactile Features and Feature Representation

One important function of the somatosensory system is the recognition of mechanical stimuli such as objects; for example, when we are searching for a key in our handbag. Tactile perception is essential for this task. In everyday life, the surfaces of objects have a panorama of stimulus properties. Each of the four classes of mechanoreceptive afferents innervating the glabrous skin differentially transmits a particular profile of stimulus features, which is signaled to the cerebral cortex, leading to perception. Thereby, different sets of somatosensory afferents are activated simultaneously, and their signals are integrated in the cortex to form a perceptual representation of the sensory input.

In the following, general features of tactile stimuli are introduced as well as their representation in the somatosensory system. The reader is referred to Mountcastle 2005 for more detailed information. In addition, Chapter 4 of the present thesis elaborates on tactile feature processing in the somatosensory system.

Orientation

Information about the orientation of objects is of importance to most primates for their life in trees but also to humans for their use of tools. The perception of orientation has been studied using intended or scanned bars, ellipsoids, cylinders, etc. with different orientations. Orientation encoding requires high innervation density and spatial acuity. These properties are only provided by SAI afferents, which signal orientation information in their population response to postcentral neurons (Dodson et al. 1998; Khalsa et al. 1998). At the cortical level, neurons in SI and SII have been found to selectively respond to specific stimulus orientations (DiCarlo and Johnson 2000; Hsiao et al. 2002).

Motion and Direction

The sensation of motion is created by sequential displacement of objects contacting the skin and involves lateral force and skin stretch. Motion and/or direction information is signaled by the excitation of different RAI populations, which successively overlap in the line of movement. SAII afferents transmit the degree and direction of skin stretch. In the cortical somatosensory system, responses of SI neurons are modulated by the direction of stimulus motion (Costanzo and Gardner 1980; Warren et al. 1986).

Texture

The texture of an object's surface can be perceived by scanning movements of the finger tip or tongue. Texture perception involves the independent dimensions of roughness, softness, and stickiness, resulting from the repetitive arrangement of small constituent parts of surfaces. A spatial neural mechanism accounts for the perception of textures with larger element separation, whereas a vibratory one serves perception of finer surface textures. SAI afferents provide a spatial variation code in their population response, and RAII responses transmit a neural representation of vibrations transmitted through textured surfaces (for review see Johnson and Hsiao 1992; Hollins et al. 2002). SAI afferent signals converge upon neurons of area 3b that encode differences in texture by differences in their discharge frequency (Darian-Smith et al. 1984; Sinclair and Burton 1991).

Spatial Form and Pattern

Form is the specific two-dimensional geometric structure of a surface or object. The capacity for form and pattern perception at the finger tip is the same whether the object is touched actively or applied passively, whether it is stationary or moving across the skin; and it is set by the resolving power of the somatosensory system. SAI responses provide a robust, isomorphic neural image of the spatial structure of objects contacting the skin (Phillips et al. 1990). Similar responses have been found for SI neurons (Phillips et al. 1988), which form the basis for form and pattern perception.

2.4. The Model of Somatosensory Processing for Perception and Action

As described in the previous section, the mechanoreceptive representation of tactile stimulus properties is transmitted to SI in the postcentral gyrus, where the input is elaborated further. The processing of simple and complex features in SI may be only the first in a series of increasingly abstract representations of mechanical stimuli. In this process, not only stimulus characteristics but also the purpose of processing may shape the way the information is being processed. According to this, Dijkerman and de Haan proposed a model of somatosensoy processing for perception and action (Figure 2.2; Dijkerman and de Haan 2007). The authors suggest that a cortical pathway projecting from SI via SII to the insula is responsible for conscious somatosensory perception and recognition of objects, with the PPC contributing to spatio-temporal integration. Information about the properties of an object is provided by higher-order association areas combining tactile features elaborated in earlier processing steps. A second cortical stream projects from SI to SII and terminates in the PPC for action-related somatosensory processing. The aforementioned idea of a ventral somatosensory pathway terminating in the insula for perceptual learning was incorporated into the model and extended to include the PPC involved in tactile object perception. The dorsal somatosensory pathway is represented by the second cortical stream suggested for action-related processing, which terminates in the PPC. According to this, the model proposes a "what" versus "how" distinction along two different processing streams for somatosensory perception and action. However, tactile perception usually requires close cooperation with action-related processes. The distinction made between the two somatosensory streams are thus less independent than the visual "what" versus "where" distinction. A further segregation can be seen between somatosensory processing of external stimuli (objects and their features) and internal targets (bodily awareness) involving proprioception, which is referred to as body image in Figure 2.2. The present thesis aims at extending this model focussing on conscious perception of external tactile input and the role of feature processing and attentional top-down modulation for tactile perception. Chapter 4 deals with the neural pathways underlying feature-specific higher-order processing of tactile stimulus attributes. Chapter 5 addresses top-down directed attentional gating of somatosensory perception.

2.5. Attention in the Somatosensory System

In a world characterized by sensory overload, we are aware of just a small portion of the sensory information that reaches the organism. As a consequence, we pay attention to some things at the expense of others. Attention plays a major role in how sensory information is being processed and perceived. Imagine a mild summer evening deep in conversation with friends you may not have noticed that you got bitten by mosquitoes

2.5. Attention in the Somatosensory System

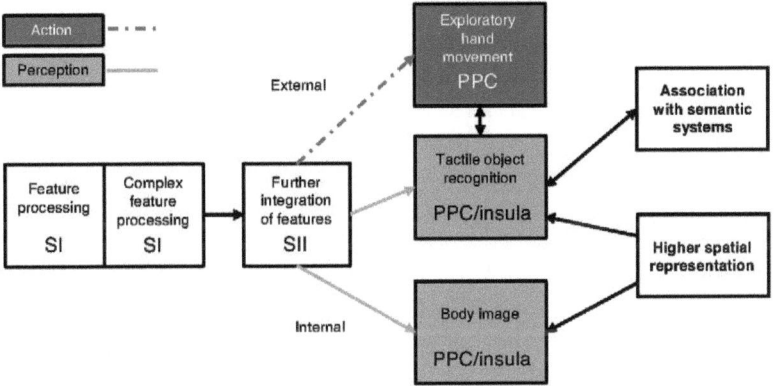

Figure 2.2.: Model of somatosensory processing for perception and action. Dark grey boxes and broken lines depict areas and pathways involved in somatosensory processing for action. Light grey boxes and lines show those involved in somatosensory processing for perceptual recognition. The model further differentiates between the processing of external stimuli and internal targets. SI, primary somatosensory cortex; SII, secondary somatosensory cortex; PPC, posterior parietal cortex. Adapted from Dijkerman and de Haan 2007.

several times. By this example the power of selective attention is illustrated indicating two important aspects of sensory processing: (1) we have a limited capacity for information processing, which requires that we overlook most of the incoming sensory input, and (2) attention may be under cognitive control, like a spotlight that can be focussed on specific sensory information. As a definition, attention is a neural mechanism that allows directing mental efforts onto specific objects or events, such as external stimuli or internal mental states (Vega-Bermudez and Hsiao 2002).

Top-Down and Bottom-Up Mechanisms of Attention

Two different processes of attention can be distinguished. As described above, attention can be controlled by cognitive factors such as knowledge, expectation, and current intentions, which allows us to focus our efforts on a specific stimulus or task. This top-down, or endogenous, mechanism provides us with the ability to selectively suppress irrelevant information and enhance the representations of sensory stimuli that are of current relevance.

Chapter 2. The Human Somatosensory System

On the other hand, sensory perception can also be dominated by external events. Unexpected, salient, and potentially dangerous events are given high priority, and are perceived at the expense of ongoing neural activity. These stimuli are processed by bottom-up, or exogenous, mechanisms, which are essential as they capture our attention to stimuli such as an alarm signal that may be of great importance to us. An overview can be found in Corbetta and Shulman 2002.

Effects of Attention

There are several ways to characterize the effects of attention. Using psychophysiological methods, differences in human sensory performance are studied while selective attention is switched between different sensory modalities, different body parts, or different stimulus properties. Human subjects are able to rapidly engage attention to task-relevant targets, which are in consequence perceived more rapidly and accurately. An example is the filtering task, in which two or more features of a stimulus can change. In this task, selective cueing to the feature that changes reduces response times and improves accuracy in detecting that change compared to a neutral or even false cue. Selective cueing focusses attentional resources (Posner 1986), which also holds for selective attention in the somatosensory system. In a previous behavioral study, changes in tactile stimulus attributes were detected with higher accuracy when validly cued compared to falsely or neutrally cued (Sinclair et al. 2000).

Another approach is to study the effects of attention directly on the responses of neurons in the nervous system, using neurophysiological recordings in monkeys or, non-invasively, neuroimaging methods in humans. The assumption is that changes in behavior are accompanied by changes in neural activity that reflect the attentional effort required to perform the task. Indeed, selective attention modifies activity specifically in those brain areas that are usually driven by the attended kind of stimulation (Corbetta et al. 1991). In the competition between different resources selective attention serves to enhance the processing capacities of particular neural populations during task performance (Desimone and Duncan 1995). In the somatosensory system attentional effects can be found in SI and SII as demonstrated by several neuroimaging studies (Mima et al. 1998; Burton et al. 1999; Johansen-Berg et al. 2000). Increased activity due to attended tactile stimulus attributes was further observed in inferior and superior parietal areas as well as in prefrontal cortex (Van Boven et al. 2005; Burton et al. 2008). Burton and colleagues reasoned that activity in frontal and parietal areas indicates resource mobilization for target or response selection. These cognitive components are involved whenever discriminating stimulation attributes, whether tactile or visual (Burton et al. 2008). In the present thesis, Chapter 5 deals with the effects of selective attention on the processing of tactile stimulus attributes.

Chapter 2. The Human Somatosensory System

Chapter 3.

Introduction to Functional Magnetic Resonance Imaging

Functional magnetic resonance imaging (fMRI) is an imaging technique used to measure the hemodynamic response, i.e., the change in blood flow, which is related to neural activity, in the human or animal brain. As one of the most recently developed neuroimaging methods, it dominates the field of functional brain mapping due to its non-invasiveness, its absence of radiation, and its wide availability.

In this chapter, the basics of fMRI are introduced, beginning with a short description of the underlying physics and the relationship between hemodynamics and magnetic resonance (MR). We move on to a general description of the methods for statistical analysis of fMRI data that were used in the present thesis. These include the general linear model for conventional statistical inference as well as psychophysiological interactions and dynamic causal modeling for the analysis of functional integration.

The interested reader is referred to Huettel et al. 2009 for a comprehensive overview of all aspects of fMRI. For more detailed information on statistical data analysis, we recommend Friston et al. 2006.

3.1. Physics of Magnetic Resonance Imaging

Hydrogen nuclei in the human body bear magnetic moments that align with the direction of a static magnetic field B_0. A radio frequency pulse causes the nuclei to alter their magnetization alignment relative to the field. The frequency associated with this "spin flip" is the Larmor frequency

$$\omega_L = gB_0 \qquad (3.1)$$

where g is the gyromagnetic ratio of hydrogen (Huettel et al. 2009). Radio frequency pulses and magnetic field gradients can be used with different timing and amplitude parameters. Adjustments to these parameters allow imaging of various physiological properties, e.g., structure (anatomical imaging), flow (perfusion imaging), or neural activity (functional imaging).

When the electromagnetic field is turned off, the excited magnetic spins return to the original alignment. The longitudinal relaxation can be described by the time constant T_1. The transverse relaxation decays with the time constant T_2. In physiological tissue the transverse relaxation is more rapid due to local field inhomogeneities, which can be described by the time constant T_2^*. Changes in the T_2^*-weighted MR signal can be used as indicator for changes in neural activity as described in the following.

3.2. Hemodynamics and Magnetic Resonance

Neural activity is closely linked to metabolic changes, which include increased consumption of energy in the respective brain regions (Siesjo 1978). Energy is supplied in the form of glucose and oxygen. Increased oxygen consumption leads to a rise in local blood volume and local blood flow in regions of increased neural activity, which is referred to as hemodynamics. The precise mechanism behind this neurovascular coupling is still not completely understood. As a consequence, the relative concentration of oxyhemoglobin and deoxyhemoglobin changes in activated brain regions in concert with changes in blood volume and blood flow. Oxygenated hemoglobin is diamagnetic, whereas deoxygenated hemoglobin has paramagnetic properties, leading to local field inhomogeneities. Thus, the T_2^*-weighted MR signal of blood differs depending on the level of oxygenation, which is referred to as blood-oxygen level-dependent (BOLD) signal (Ogawa et al. 1990). Activated brain regions that are relatively low on deoxyhemoglobin show a slower decay of the T_2^*-weighted MR signal than non-activated brain regions do, which results in an increased BOLD signal. In brief, fMRI measures the relative absence of deoxyhemoglobin in a given brain region, which is an indicator for local neural activity (for review see Logothetis and Wandell 2004). Evoked by sensory stimulation, the hemodynamic response begins to rise at about 2 s post-stimulus and peaks at 5 to 9 s after stimulus onset before falling back to baseline. The measured difference in signal intensity is very small, but given many repetitions of an experimental manipulation, statistical methods can be used to determine those brain areas that reliably show this difference more frequently, and therefore may be related to that experimental manipulation.

3.3. Statistical Analysis of Functional Imaging Data

Before any statistical analysis takes place, functional imaging data generally undergo a series of spatial transformations in order to reduce variance in the time-series, which may be induced by head movement or anatomical differences among subjects. The statistical analysis of fMRI data used in the present thesis is voxel-based, that is, the brain is subdivided into voxels. Voxels are cubes of variable volume depending on scanning resolution (typically between 8 and 27 mm^3) that represent the activity of a particular coordinate in three-dimensional space. Such a voxel-based analysis assumes that all points in a specific voxel time-series derive from the same site in the brain. Violations of this assumption such as subjects' head movement may induce artifactual changes in the voxel values. In the following, the most common transformations are briefly introduced. The preprocessing procedures as well as the statistical approach used for the individual data analyses in the present thesis are described in Sections 4.2 and 5.2.

3.3.1. Image Preprocessing

The first preprocessing step is usually the realignment or motion correction of the images, which accounts for subjects' movement of the head between scans. Using a least-squares approach and an affine 6-parameter (rigid body) transformation, the time-series are adjusted such that each of the voxels corresponds approximately to the same site in the brain.

Functional neuroimaging studies usually involve several participants with slightly differently shaped brains due to differences in overall brain size or variations in the topography of gyri and sulci of the cerebral cortex. For that reason the data are transformed using linear or nonlinear warps into a standard anatomical space such as the Talairach space (Talairach and Tournoux 1988) or Montreal Neurological Institute (MNI) brain templates. This process is referred to as spatial normalization. Even single-subject analyses (e.g., case studies) proceed in a standard anatomical space in order to enable the comparison of the results of different studies.

Finally, the data are often spatially smoothed using Gaussian filters varying in size in order to account for effects due to residual differences in functional and gyral anatomy.

3.3.2. Statistical Analysis

Statistical analysis of imaging data comprises (1) modeling the data in order to separate task-related from confounding effects and residual variability and (2) making inferences about regionally specific task-related effects (in relation to the error variance). This classical inference is a statistical comparison of effects due to the experimental manipulation with the error variance employing Student's T or F statistics. This kind of spatially extended statistical approach to test hypotheses about regionally specific effects is referred to as Statistical Parametric Mapping (Friston et al. 1991).

The modeling of fMRI data uses a mass-univariate approach based on the general linear model (GLM), also known as multiple regression analysis. The GLM can be described by the equation

$$Y = X\beta + \epsilon \qquad (3.2)$$

that expresses the observed response variable Y in terms of a linear combination of explanatory variables X and an additional error term ϵ (Friston et al. 1995). The so-called design matrix X that contains the explanatory variables defines the experimental design. It has one row for each scan and one column for each effect that may be task-related or that may confound the results (explanatory variables, covariates, or regressors) as schematically illustrated in Figure 3.1 (on the left). The explanatory variables are modeled by convolving a series of stick or box functions (indicating the onset of events or epochs, respectively) with a set of basis functions. This set of basis functions is referred to as hemodynamic response function (HRF), which models the hemodynamic convolution that the brain applies to the input.

After the model is fitted to the experimental data, inferences about relative contributions of each of the explanatory variables can be made. This is done using T or F contrast vectors to produce statistical parametric maps (SPMs; Figure 3.1, right). Contrast vectors are referred to as contrasts and contain contrast weights that yield a weighted sum or compound of parameters tested. As an example, the contrast [1 1 -1 -1 ...] displayed above the columns of the first-level design matrix in Figure 3.1 can be used to compare experimental conditions. The SPM contains T or F statistics that allow delineating the relative statistical significance of regionally specific effects. Inference is usually based on height and/or spatial extent thresholds at the peak or cluster level. Peak level refers to the chance of finding a peak with this or a greater height, and cluster level refers to the chance of finding a cluster with this many or a greater number of voxels.

3.3. Statistical Analysis of Functional Imaging Data

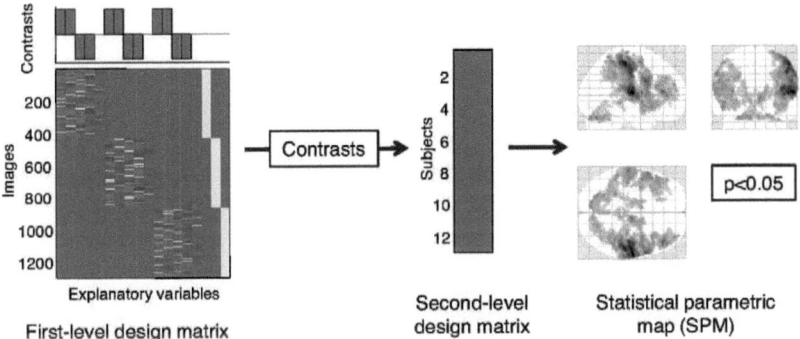

Figure 3.1.: Mixed-effects analysis for population inference. The analysis of group data proceeds in two stages using models at two levels. The exemplary first-level design matrix of the within-subject analysis includes five explanatory variables (onsets of experimental conditions convolved with a hemodynamic response function) and a constant term per session. Subjects' contrast images of the first-level analysis enter the second-level design matrix. Effects are tested using T or F statistics that constitute the SPM. A maximum intensity projection of the thresholded SPM that conforms to a standard anatomical space is shown on the right.

The analysis of data from group studies usually proceeds in two stages using models at two levels (Figure 3.1). This approach is termed mixed-effects analysis (Friston et al. 1995). At the first level, within-subject analyses are implemented for each subject (Figure 3.1, left), which are also used for the analysis of case studies. In a group study, however, one wishes to make inferences about the population, from which the subjects were drawn. For this purpose, contrast images from each subject's first-level analysis are used as summary measures of subject responses, which are then entered as data into a second-level model (Figure 3.1, middle). The examplary second-level design matrix in Figure 3.1 contains one observation (i.e., contrast) per subject. The error variance is computed using the between-subjects variability of estimates from the first-level analysis. Inference about the population can then be made by entering contrasts into the second-level analysis.

As mentioned above, the analysis of imaging data is a voxel-based approach, which means that many statistical tests are being conducted. The comparison of signal intensities at a large number of voxels may result in falsely detecting background brain activity as task-related activity. In order to prevent false positives, a correction for multiple dependent comparisons has to be made, which can be achieved using Gaussian random field

(GRF) theory (Worsley et al. 1996). GRF theory allows for the adjustment of the p-value for the search volume of the SPM. This procedure for continuous data such as images is equivalent to the Bonferroni correction for discrete data.

3.4. Functional Integration and Effective Connectivity

Up to here, we only considered the identification of regionally specific brain activity that can be attributed to a specific task or experimental manipulation, which is referred to as functional specialization. Another principle of brain organization is functional integration, which concerns interactions among specialized brain areas and how these interactions depend on the given context. Functional integration can be assessed by testing the correlations among activity in different brain areas (i.e., functional connectivity) or by trying to explain the activity in one area in causal relation to other areas (i.e., effective connectivity). By definition, effective connectivity refers explicitly to the influence that one neural system exerts over another (Friston 1995). The fundamental difference between functional and effective connectivity is that correlations between activity in different brain areas do not necessarily imply direct or indirect neuronal interactions, whereas effective connectivity does. In the following, the principles of the two approaches to estimate effective connectivity used in the present thesis are introduced. These are psychophysiological interaction (PPI) models and dynamic causal modeling (DCM). More detailed information can be found in Friston et al. 2006.

3.4.1. Psychophysiological Interaction Models

The concept of PPIs was introduced as a method for the analysis of effective connectivity in the late nineties (Friston et al. 1997). According to this, activity in any given brain area can be explained in terms of an interaction between the influence of another area (i.e., the physiological variable) and some experimental factor (i.e., the psychological variable). PPIs combine two important concepts, namely (1) factorial designs, which are used to investigate interactions between two or more psychological variables, and (2) effective connectivity, which refers to interactions between two physiological variables. Using a relatively simple procedure PPI models provide evidence for interactions between distributed cortical networks and enable inferences about task-dependent changes in cortical organization.

3.4. Functional Integration and Effective Connectivity

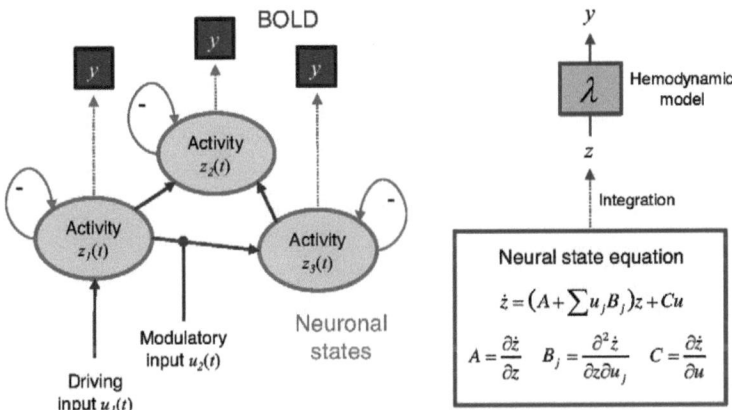

Figure 3.2.: Conceptual basis of DCM. The neural dynamics in a cortical network (left) is modeled using a bilinear state equation (box), which is integrated to give the predicted neural dynamics (z). z enters a model of the hemodynamic response (λ) to give predicted BOLD responses (y). Model parameters are estimated such that the difference between predicted and measured BOLD responses is minimized. Experimental perturbations enter the model in form of driving inputs (u_1) that elicit responses directly, and modulatory inputs (u_2) that change the connection strengths. Intra-areal inhibition prevents self-reinforcing activity.

Technically, PPI terms are introduced as explanatory variables into the GLM, which allows statistical parametric mapping to estimate significance and magnitude of the effect. Equation 3.2, which describes the GLM, is augmented as follows to include the PPI term (Friston et al. 2006):

$$Y = [X \; u \times y_i]\beta + \epsilon \qquad (3.3)$$

The Hadamard product $u \times y_i$, which represents the PPI, is obtained by multiplying the psychological variable u (the input) and the physiological variable y_i (the response) measured at the i-th brain region. The design matrix X, which contains the main effects, has to be included to allow assessing additional explanatory power of the PPI. The regional significance of the PPI can be estimated using an SPM as is the case with a traditional GLM analysis.

Chapter 3. Introduction to Functional Magnetic Resonance Imaging

3.4.2. Dynamic Causal Modeling

DCM, on the other hand, is not a method for making inferences about BOLD responses but rather directly about the neural processes that underlie the measured BOLD time series. Introduced by Friston et al. 2003, the idea is to estimate the parameters of a neural system model such that the difference between the predicted BOLD responses and the observed BOLD time series is minimized (see Figure 3.2 for an overview). A prediction of the BOLD signal is obtained by transforming the modeled neural dynamics into hemodynamic responses. Within this framework, an fMRI experiment can be considered as experimental manipulation of interactions among brain regions via effective connectivity. Incidentally, the use of DCM is not restricted to functional imaging data. DCMs can theoretically be formulated for any measurement technique using appropriate state equations and observation models.

Neural State Equations

Importantly, the neural dynamics in a system of k interacting brain regions can not directly be observed using fMRI. The neural dynamics is modeled at the hidden level of a DCM. Each modeled region i is represented by a state variable z_i, and the dynamics of the system is described by the change of the neural state vector over time. For k neuronal states $z = [z_1, \ldots, z_k]^T$, the temporal evolution is formulated in the neural state equation

$$\dot{z} = F(z, u, \theta^n) \tag{3.4}$$

as a function of the current state, the external inputs u, and the parameter vector θ^n that defines the functional architecture and interactions among brain regions at a neuronal level (Friston et al. 2006). The inputs may enter the model in one of two forms: (1) driving inputs (e.g., sensory stimulation) elicit responses through direct influences on early sensory regions and (2) modulatory inputs (e.g., changes in cognitive set) may change the coupling strength among connected regions. z and u are time-dependent whereas θ^n is time-invariant. In DCM for fMRI, F has the bilinear form

$$\dot{z} = (A + \sum u_j B_j)z + Cu \tag{3.5}$$

3.4. Functional Integration and Effective Connectivity

The parameters $\theta^n = \{A, B_1, \ldots, B_l, C\}$ can be expressed as partial derivatives of F:

$$A = \frac{\partial F}{\partial z} = \frac{\partial \dot{z}}{\partial z}$$
$$B_j = \frac{\partial^2 F}{\partial z \partial u_j} = \frac{\partial A}{\partial u_j} \quad (3.6)$$
$$C = \frac{\partial F}{\partial u}$$

These coupling matrices describe the causal components, which underlie the modeled neural dynamics. The connectivity matrix $A(k,k)$ represents the context-independent effective connectivity among k brain regions mediated by anatomical connections. This can be thought of as the regional coupling in the absence of external input. The matrices $B_j(k,k)$ encode the context-dependent changes in effective connectivity induced by the j-th input u_j. These terms are referred to as bilinear because they are second-order derivatives. The matrix $C(k,l)$ gives the external inputs into the system that drive regional activity. The posterior densities of these parameters allow for inferences about the impact that experimental perturbations may have on the dynamics in the modeled system.

The Hemodynamic Model

In order to predict the BOLD responses, the model of the neural dynamics is combined with a hemodynamic model, the so-called Balloon model (Friston et al. 2000). Briefly summarized, it consists of four biophysical state variables (s, f, v, q), which form the BOLD signal and transform neuronal activity into hemodynamic responses. A set of differential equations describes the relations between these state variables using five parameters $\theta^h = \{\kappa, \gamma, \tau, \alpha, \rho\}$. The activity-dependent vasodilatory signal s leads to increases in blood flow f and subsequently to changes in blood volume v and deoxyhemoglobin content q. The predicted BOLD signal y is a non-linear function of v and q. Please refer to Friston et al. 2000 for more detailed information.

The Likelihood Model

Combining the neural and hemodynamic states into a joint state vector $x = \{z, s, f, v, q\}$ gives the full forward model

$$\dot{x} = F(x, u, \theta)$$
$$y = \lambda(x) \quad (3.7)$$

Chapter 3. Introduction to Functional Magnetic Resonance Imaging

with the joint parameter vector $\theta = \{\theta^n, \theta^h\}$. For any given set of parameters θ and inputs u, predicted BOLD responses $h(u, \theta)$ can be obtained by integrating the joint state equation and passing it through the output non-linearity λ

$$y = h(u, \theta) + X\beta + \epsilon \qquad (3.8)$$

Additional observation error ϵ and any other confounding effects X (e.g., scanner-related signal drifts) transform the dynamic model into a likelihood model. The vector β contains the unknown coefficients for confounds.

Estimation and Inference

Equation 3.8 is the basis for the estimation of the neural and hemodynamic parameters from the measured BOLD responses using a Bayesian approach. Details of the parameter estimation scheme can be found in Friston et al. 2003. After fitting the model to the fMRI data, the posterior distributions of the parameters can be used to test hypotheses about significance and magnitude of effects at the neural level. In principle, any of the parameters in the model could be used for making inferences, however hypothesis testing usually aims at context-dependent changes in effective connectivity (i.e., parameters from the B matrices). This allows making inferences about the coupling among the modeled brain areas and how the coupling is influenced by changes in the experimental context.

Bayesian Model Selection (BMS)

Given some data and different models, how do we know, which of these models is optimal? This question of model goodness concerns any kind of modeling approach in general. Importantly, it is not sufficient to consider the relative fit of the different models, one also needs to take into account their relative complexity. There is a trade-off between model fit and generalizability: model fit increases with increasing model complexity but its generalizability decreases, which is referred to as overfitting.

Therefore, it is important not to select the optimal model but the one with the best balance between fit and complexity. In a Bayesian context, this can be approached by comparing their model evidence $p(y|m)$. According to Bayes theorem

$$p(\theta|y, m) = \frac{p(y|\theta, m)p(\theta|m)}{p(y|m)} \qquad (3.9)$$

the evidence of a model m is simply the normalization term for the product of the likelihood of the data y and the prior probability of the parameters θ, given by

$$p(y|m) = \int p(y|\theta, m)p(\theta|m)d\theta \qquad (3.10)$$

3.4. Functional Integration and Effective Connectivity

For non-linear models, this integral can not be solved analytically but there are various approximations, e.g., the Laplace approximation used by DCM (Penny et al. 2004), which takes into account that the optimal model should represent the best compromise between fit and complexity.

Two models m_i and m_j can then be compared using the Bayes factor:

$$B_{ij} = \frac{p(y|m_i)}{p(y|m_j)} \quad (3.11)$$

When $B_{ij} > 1$ the data favor model m_i over model m_j, when $B_{ij} < 1$ the data favor m_j. A Bayes factor of greater than 20 is strong evidence for the given model (Kass and Raftery 1995). This corresponds to the posterior probability of greater than 0.95 for this model (similar to the p<0.05 criterion employed in classical inference). Note that this comparison is only valid for identical data y in all models. In DCM for fMRI, the data represent the observed BOLD time series of all areas in the model. Therefore, only models containing the same areas can be compared.

BMS at the Group Level

For BMS at the group level, one has to decide between fixed-effects (FFX) and random-effects (RFX) analysis. In the FFX case, it is assumed that the modeled processes do not vary across the subjects in the population, which applies, for instance, when studying basic physiological mechanisms. Under the FFX assumption the product of the individual Bayes factors can be used as criterion for the comparison (Stephan et al. 2009). This group Bayes factor quantifies the relative goodness of models considering the group as a whole.

However, when investigating cognitive tasks that may be performed using various cognitive strategies, it is more appropriate to apply an RFX BMS procedure. Using variational Bayes the posterior probabilities of competing models are estimated (Stephan et al. 2009). This allows assessing how likely it is that a specific model generated the data of a randomly chosen subject (i.e., the expected posterior model probability). A further measure that can be assessed is the exceedance probability that one model is more likely than any other model, given the group data.

Chapter 3. Introduction to Functional Magnetic Resonance Imaging

Chapter 4.

Feature-Specific Processing of Tactile Stimulus Attributes

As introduced in Chapter 2, human somatosensation can supply the organism with information about "where" (e.g., on the left forearm), "what" (e.g., a raindrop or an insect), and "how" (e.g., moving towards the hand) environmental stimuli are experienced. Compared to vision, however, the neuronal pathways underlying the processing of specific tactile stimulus attributes are still largely controversial.

In this chapter, we present the experimental investigation carried out to study the neuronal pathways underlying feature-specific processing of tactile stimulus attributes, namely motion and pattern. This is the first of the two studies that form the basis for the present thesis. First, we introduce the latest state of the knowledge in the field and give a short motivation for this investigation. We move on to a detailed description of the methods used as well as a comprehensive presentation of the results. Finally, the significance of the results is discussed and how the conclusions drawn contribute to the current knowledge in the field.

4.1. Introduction

The best studied dimension of somatosensory perception is the location of tactile stimuli on the body surface, which is long known to be represented in a somatotopic manner in SI and SII, as described in Section 2.2 of the present thesis. There is, however, accumulating evidence that also more complex tactile stimulus attributes, such as motion and pattern, are already coded in these areas of the somatosensory processing circuits. For instance, neurophysiological studies in monkeys have identified populations of SI neurons whose

Chapter 4. Feature-Specific Processing of Tactile Stimulus Attributes

responses are modulated by the direction of stimulus motion (Costanzo and Gardner 1980; Warren et al. 1986; Ruiz et al. 1995). More recently, orientation-tuned neurons have been found in SI (DiCarlo and Johnson 2000; Bensmaia et al. 2008) and SII (Hsiao et al. 2002; Fitzgerald et al. 2006), and SI has been shown to play an important role in tactile pattern recognition (Phillips et al. 1988; Hsiao et al. 1993; DiCarlo and Johnson 2002). Human neuroimaging studies have consistently demonstrated activity in SI and SII related to the discrimination of moving tactile stimuli (Burton et al. 1999; Bodegård et al. 2000), and SI has been associated with the processing of tactile form (Bodegård et al. 2001; Van Boven et al. 2005). The involvement of SII in tactile form discrimination is not yet fully elucidated as there is both supporting (Van Boven et al. 2005) and contradicting evidence (Hinkley et al. 2009).

Beyond SI and SII, the pathways of tactile motion and pattern processing are less clear. There is some evidence that the anterior part of the supramarginal gyrus in the IPC is involved in tactile discrimination of shapes and/or form (Hadjikhani and Roland 1998; Bodegård et al. 2001; Van Boven et al. 2005). Lesion studies further indicated that there are somatosensory association areas in the IPC assumed to be specific to tactile shape processing (Reed and Caselli 1994; Reed et al. 1996).

Functional imaging studies have shown that the processing of tactile stimulus features is often characterized by activity in areas that are traditionally associated with visual equivalents of these features. Tactile motion, for example, has been found to engage area hMT+/V5 in the middle temporal cortex, both in sighted (Hagen et al. 2002; Blake et al. 2004; Summers et al. 2009) and in congenitally blind individuals (Ricciardi et al. 2007; Matteau et al. 2010; Sani et al. 2010). First identified as responsive to visual motion in the middle temporal cortex of the monkey (Dubner and Zeki 1971), area MT/V5 and neighboring motion-sensitive areas such as the medial superior temporal area (MST) were collectively termed MT+/V5. The human homologue of this region identified using non-invasive neuroimaging (Zeki et al. 1991; Watson et al. 1993; Tootell et al. 1995) has long been considered a purely visual motion-sensitive area. Similarly, processing of tactile shapes typically activates extrastriate areas such as the LOC, which plays a crucial role in visual object shape perception (Amedi et al. 2001; Stoesz et al. 2003; Pietrini et al. 2004; Stilla and Sathian 2008). One possible explanation for the recruitment of these specialized visual areas during processing of complex tactile information may be visual imagery of tactile stimulus features (e.g., Lacey et al. 2010; but see Beauchamp et al.

2007). The actual function of these areas during tactile information processing, however, remains poorly understood.

Recent findings suggest that somatosensory processing of tactile motion and pattern may also involve areas that are not directly associated with any specific sensory modality. One such multisensory region is the intraparietal area lining the IPS in the posterior parietal cortex. The IPS contains multiple parietal fields, among others the anterior and the ventral intraparietal area (AIP and VIP) described in monkeys (Culham and Kanwisher 2001; Grefkes and Fink 2005; Culham et al. 2006). Bremmer and colleagues identified a ventral area in the human anterior IPS (aIPS) involved in visual, tactile, and auditory motion processing (Bremmer et al. 2001) that might be equivalent to the motion-sensitive multisensory association area VIP in monkeys. Similar to the monkey AIP, which is highly responsive to size, shape, and orientation of objects, parts of the human aIPS were shown to be engaged in visual and tactile discrimination of grating orientation (Shikata et al. 2001; Kitada et al. 2006), in form perception (Bodegård et al. 2001; Van Boven et al. 2005), and in object recognition (Binkofski et al. 1999; Grefkes et al. 2002).

Despite these recent advances in delineating the cerebral networks engaged in tactile form and motion processing, a clear consensus regarding the specific processing pathways is still lacking. This may in particular be due to large methodological differences regarding stimulus characteristics (two- or three-dimensional stimuli), task (discrimination, recognition, or naming), and exploratory strategies (passive or active, single digit or whole hand). In fact, compared to studies of vision, experimental investigation of complex tactile sensations is often complicated by the problem of mechanically administering well-described and replicable cutaneous input that creates the percept of interest (such as motion, shape, or object orientation), ideally unconfounded by active motor exploration.

Here, we investigated the processing of complex tactile information using a finger-tipsized multi-pin stimulation device similar to a Braille display to induce the sensation of moving and stationary bar patterns within a circumscribed area of glabrous skin. The percepts of interest were created during passive touch under fully specified physical stimulus conditions, thereby achieving a high level of control over the mechanical input. Using a passive stimulation paradigm, not requiring any overt response to the stimuli of interest, further ensured that the results were not affected by response-induced BOLD signal changes in the somatosensory system. Both for moving and for stationary patterns, matched control stimuli were designed that preserved the overall physical dynamics of

Chapter 4. Feature-Specific Processing of Tactile Stimulus Attributes

the stimuli of interest but did not induce a percept of motion or pattern, allowing us to contrast motion- and pattern-specific activity in a balanced experimental design.

For a fine-grained analysis of the neuronal networks engaged in processing of tactile motion and patterns, we utilized ultra-high-field fMRI at 7 Tesla. In the analysis, we investigated to what extent specialized visual and/or multisensory areas are recruited under tightly controlled tactile stimulation conditions, and sought to determine the relations between subject-specific brain activity in these areas and the outcome of individual behavioral performance in identifying the stimulus attributes of interest. Furthermore, we studied stimulus-induced changes in effective connectivity between these specialized cortical areas and somatosensory cortices in order to characterize functional integration during tactile information processing.

4.2. Materials and Methods

Participants

Thirteen healthy volunteers (aged between 22 and 35 years; nine males, one left-handed) participated in the study with written informed consent. The study corresponded to the Human Subjects Guidelines of the Declaration of Helsinki and was approved by the Local Ethics Committee at the faculty of medicine, Otto-von-Guericke-University of Magdeburg.

Stimuli

Tactile stimulation was applied to the left index finger by a 16-dot piezoelectric Braille-like display (4x4 quadratic matrix, 2.5 mm spacing) controlled by a programmable stimulation device (Piezostimulator, QuaeroSys, St. Johann, Germany). On each trial, the pins of the display were driven for 4000 ms by a 144 Hz sinusoidal carrier signal, which was amplitude-modulated by different sets of rectified 2 Hz sine functions (see Figure 4.1a for illustration). Four different stimulus types were designed to create the sensation of (1) a moving bar pattern, (2) a moving random stimulus, (3) a stationary bar pattern, or (4) a stationary random stimulus.

For the moving bar pattern (1), all pins forming a diagonal on the quadratic display were driven by a rectified 2 Hz sine function, with half cycles repeating every 250 ms. From diagonal to diagonal, the phase of the sine function was shifted by $\pi/4$ (62.5 ms, corresponding to a quarter of its rectified half cycle; see Figure 4.1b for illustration). This directed propagation of the diagonals' phase across the display plane created the sensation

4.2. Materials and Methods

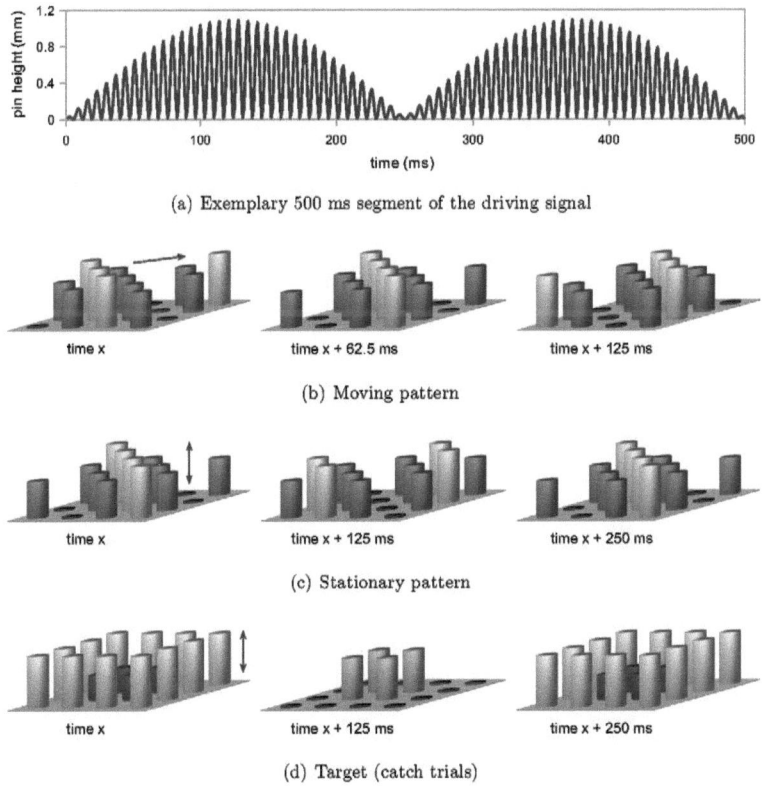

(a) Exemplary 500 ms segment of the driving signal

(b) Moving pattern

(c) Stationary pattern

(d) Target (catch trials)

Figure 4.1.: Illustration of tactile stimuli used. (a) The pins' driving signal was a 144 Hz sinusoidal carrier, which was amplitude-modulated by a rectified 2 Hz sine function. (b) A directed propagation of the diagonals' sine phase across the display plane resulted in a percept of a bar pattern travelling smoothly across the finger tip. Both upward and downward diagonal orientations were used corresponding to orthogonal moving directions. For moving random stimuli, each of the four sine phases was assigned to four randomly chosen pins (not shown). (c) Diagonals oscillating at opposite phases created the percept of a stationary bar pattern, which was periodically elevated and retracted. Again, both diagonal orientations were used. For stationary random stimuli, the driving signals were assigned to sets of randomly chosen pins (not shown). (d) The target stimulus presented in "catch" trials was an oscillating square.

Chapter 4. Feature-Specific Processing of Tactile Stimulus Attributes

of a bar pattern travelling smoothly across the finger tip. The orientation of the diagonals and the direction of movement were randomly varied from trial to trial. For the moving random stimulus (2), the same set of driving signals was used as for (1), but each of the four different sine phases was randomly assigned to four randomly chosen pins. This corresponded to a random spatial permutation of the individual pin movements displayed in (1) and created a percept of unsystematic, disorderly motion across the display plane.

For the stationary bar pattern (3), every second diagonal of the display was driven identically to (1), thus oscillating at opposite phases of the half cycle. The remaining pins, however, (i.e., the interleaved diagonals) were all synchronously driven by the same signal (Figure 4.1c). As a result, due to the absence of a directed phase shift, the bar pattern did not propagate across the display plane but created the percept of a stationary pattern, which was periodically elevated and retracted (i.e., along the z-axis). In order to ensure identical root mean square (RMS) amplitudes in each stimulus condition at each time point, the instantaneous amplitude of the interleaved diagonals in (3) was set to the average of the corresponding driving signals used in (1). For the stationary random stimulus (4), the driving signals used for (3) were randomly assigned to randomly chosen pins, creating a percept of a disorderly structured surface, which was periodically elevated and retracted. The four stimulus types were matched according to the overall physical dynamics rather than to the subjectively perceived salience of the different stimulus attributes. This allowed for *a priori* unbiased investigation of covariations between the fMRI results and subjects' individual perceptual performance.

During scanning, presentation of the four stimulus types was not associated with any behavioral task. To ensure that participants maintained attention to the tactile stimulation throughout the scanning sessions, they were instructed to report the occurrence of a distinct tactile stimulus, which was infrequently presented ("catch trials"). The target stimulus was markedly different from the moving or stationary gratings of interest. It consisted of a square (Figure 4.1d), with the outer pins of the display oscillating at rectified 2 Hz. The inner pins of the display were driven such that across the stimulation period the average RMS amplitude of the target stimulus was identical to the average RMS amplitude of the four main stimuli. Stimulus presentation was controlled using custom MATLAB code (v7.5, The MathWorks, Natick, MA) and the Cogent 2000 toolbox (http://www.vislab.ucl.ac.uk/cogent.php).

Behavioral Data

Prior to the experiment, participants were familiarized with the stimulation device and with the different types of stimuli. Before and after scanning, subjective discriminability of the stimuli was assessed using a short behavioral classification task (pre- and post-test; 48 trials each). To avoid ceiling effects, the four main stimulus types were presented in a four-alternative classification task, which was more difficult than detecting the presence of the two stimulus features of interest (motion and pattern). Individual perceptual performance levels for motion and pattern identification were inferred from the four-alternative classification data by collapsing correct classifications of the stimulus attribute of interest while disregarding false classifications along the other stimulus dimension.

Design & Procedure

The fMRI experiment consisted of three sessions. Each session comprised 64 stimuli (16 of each stimulus type: moving bar patterns, stationary bar patterns, moving random stimuli, and stationary random stimuli), 16 null events, as well as 4 catch trials, in which the target stimulus was presented. All trials, including null events, had a duration of 4000 ms and were presented in pseudo-random serial order, such that each type of event occurred equally often in each quarter of the session. The inter-stimulus interval (ISI) was randomly varied between 2500 and 7500 ms. Participants were instructed to keep their eyes closed throughout the experiment and to press a response key with their right index finger only when they detected the target stimulus, which was presented infrequently.

fMRI Data Acquisition

Functional imaging was performed on a 7 Tesla Magnetom MRI scanner (Siemens, Erlangen, Germany) with a 24-channel head-coil system (Nova Medical, Wilmington, MA, USA). T_2*-weighted functional images were acquired using an echo planar imaging (EPI) sequence (TR = 2500 ms, TE = 21 ms, flip angle = 80°). For each session, 340 EPI volumes were obtained, each consisting of 46 axial slices covering the whole brain in an ascending order (slice thickness 2.5 mm, distance factor 0.25, in-plane resolution 2x2 mm, matrix size 106x106). To achieve this high spatial resolution with single-shot EPI acquisition, parallel imaging with an acceleration factor of two and a partial Fourier acquisition scheme (75%) were applied.

Data Preprocessing

Functional images were preprocessed and analyzed using Statistical Parametric Mapping (SPM8, Wellcome Department of Imaging Neuroscience, University College London, UK).

Chapter 4. Feature-Specific Processing of Tactile Stimulus Attributes

The first four volumes of each experimental session were discarded in order to allow the MR signal to reach equilibrium. To minimize movement-induced image distortions, each data set was realigned to the first image of the first session as described in Section 3.3. The realigned images were spatially normalized to the standard MNI template brain and smoothed using an isotropic, three-dimensional Gaussian kernel of 2 mm full-width at half-maximum (FWHM). This small kernel size was chosen to do not blur the fine-grained spatial resolution of the 7 Tesla data. To remove global effects from the fMRI time series, detrending was applied based on a voxel-level linear model of global signal (LMGS; Macey et al. 2004). The images were high-pass filtered (cut-off frequency 1/128 Hz) in order to remove low-frequency signal drifts. To reduce high-frequency noise, serial correlations were modeled using an autoregressive model.

Statistical Data Analysis

A standard two-level mixed-effects model (Friston et al. 1995) as introduced in Section 3.3 was used for statistical analysis. At the first level, multiple regression within the framework of the GLM was employed to implement a within-subject analysis. For each data set, BOLD responses were modeled by stick functions (multiplied by the stimulus duration) indicating the onsets of the stimuli of interest (i.e., moving bar patterns, moving random stimuli, stationary bar patterns, stationary random stimuli). These regressors were then convolved with a standard HRF and included in the GLM. Two further regressors were added as modulatory effects, indicating the orientation (upward or downward diagonal) of moving and stationary patterns. Null events were modeled with a separate regressor to obtain a baseline. A stimulus function for nuisance effects comprised catch trials and false alarms. To account for occasional signal intensity changes within slices due to increased susceptibility-induced frequency variations at higher field strengths (van der Zwaag et al. 2009), seven additional nuisance regressors were included. These corresponded to the first seven eigenvariates, which were extracted exclusively from signals outside the brain in a previous SPM analysis (thresholded at $p<0.05$). After the model was fitted to the experimental data, contrast images were generated from the stimulus functions' parameter estimates for each of the four stimulus types of interest. At the second level, the individual subjects' contrast images were entered into a 2x2 within-subjects ANOVA with factors motion (moving or stationary) and pattern (patterned or random). This allowed computing differential effects between the different stimulus types, using contrast vectors to produce SPMs. An additional covariate comprised the individual subjects' accuracy

4.2. Materials and Methods

levels for the classification of the four stimulus types (mean of pre- and post-test) to account for variance due to performance differences.

To identify the overall neuronal network involved in somatosensory perception, contrast images were generated at the subject level in order to compare all tactile stimulation conditions with the null events. At the group level, a one-sample t-test was calculated using the individual subjects' contrast images.

To assess possible BOLD differences for motion direction and pattern orientation, the individual subjects' contrast images that were generated from the two regressors indicating bar orientation for moving patterns and stationary patterns were entered into another ANOVA. To consider effects of both directions or orientations F-contrasts were computed. These contrasts were examined within the activation maps that resulted from contrasting moving with stationary stimuli and patterned with random stimuli to identify differential effects within motion- and pattern-specific areas.

Another matter of interest was to investigate performance-dependent covariation in regions involved in tactile motion and pattern processing. The individual subjects' accuracy in pattern and motion discrimination was correlated with the individual subjects' parameter estimates of peak voxels in the areas identified for motion and pattern processing as desribed above (hMT+/V5; $x = -44$, $y = -70$, $z = -2$ and IPC; $x = -60$, $y = -56$, $z = 32$). Subjects' accuracy in identifying moving and patterned stimuli correctly was inferred from the mean classification performance in the behavioral pre- and post-test. These tests were Bonferroni corrected for multiple comparisons.

PPI Analysis

The effective connectivity of areas involved in motion and pattern processing was assessed using PPI analyses (Friston et al. 1997) introduced in Section 3.4. Spheres with a radius of 5 mm constructed around peak voxels of the motion and pattern activation located in left hMT+/V5 ($x = -44$, $y = -70$, $z = -2$) and left IPC ($x = -60$, $y = -56$, $z = 32$) served as seed regions for extracting the first eigenvariate of the signal. At the subject level, the physiological variable was extracted and psychophysiological interaction terms were created for moving vs. stationary stimuli as well as for patterned vs. random stimuli. Subsequently, these terms were entered into GLMs. At the group level, contrast images of the PPIs were analyzed using one-sample t-tests.

Statistical Inference

All reported coordinates correspond to the anatomical MNI space. The SPM anatomy toolbox (Eickhoff et al. 2005) was used to establish cytoarchitectonic reference. To in-

Chapter 4. Feature-Specific Processing of Tactile Stimulus Attributes

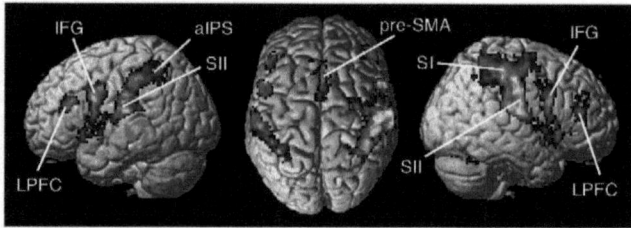

Figure 4.2.: Overall neuronal network associated with tactile stimulation. Contrasting tactile stimulation trials with null events revealed a distributed network involved in tactile information processing. SI, primary somatosensory cortex; SII, secondary somatosensory cortex; aIPS, anterior intraparietal sulcus; IFG, inferior frontal gyrus; LPFC, lateral prefrontal cortex; pre-SMA, pre-supplementary motor area.

vestigate the overall effects of tactile stimulation, we used a significance threshold of $p_{cluster} < 0.05$, corrected for multiple comparisons. This activation map was used as a mask to investigate stimulus-specific differences within the network associated with tactile stimulation. Based on *a priori* assumptions (e.g., Bodegård et al. 2000, 2001), the significance threshold was chosen more liberally ($p < 0.005$, uncorrected). This threshold was also used for the additional analysis of BOLD signal changes for motion direction and pattern orientation. To assess differential effects between stimulus conditions regarding the entire brain, that is, also including areas that may not have been generally activated by tactile stimulation, we used solely the conservative threshold of $p_{cluster} < 0.05$, corrected for multiple comparisons. This threshold was also used for the PPI analyses.

4.3. Results

Behavioral Data

On average, participants' perceptual performance was 71% (SEM = ±3%) in discriminating moving from stationary stimuli and 72% (SEM = ±3%) in discriminating patterned from random stimuli. These accuracies were inferred from the mean classification performance before and after scanning and were significantly higher than chance level (50%; both p's < 0.001, two-tailed t-test). No significant differences between the performance levels in the pre- and post-tests were observed (p = 0.43 and p = 0.18, respectively).

Region	Hemisphere	x	y	z	T-value
Primary somatosensory cortex	R	52	-22	44	12.61
Secondary somatosensory cortex	R	52	-16	16	10.29
	L	-58	-20	16	10.30
Anterior intraparietal sulcus	R	40	-48	62	8.85
	L	-38	-50	48	6.55
Inferior frontal gyrus	R	56	10	22	9.79
	L	-52	12	30	11.63
Lateral prefrontal cortex	R	42	40	16	5.81
	L	-42	32	18	7.59
Pre-supplementary motor area	R/L	2	16	46	7.56
Insular cortex	R	36	22	-2	9.25
	L	-36	18	-2	6.02
Thalamus	R	10	-14	4	5.81
	L	-10	-12	-2	5.38
Cerebellum	R	24	-66	-24	8.53
	L	-24	-50	-28	9.95

Table 4.1.: Functional regions active during tactile stimulation. x, y, z are MNI coordinates (mm). T-values are local maxima within a significant cluster of activated voxels with $p_{cluster}<0.05$, corrected for multiple comparisons (group-level analysis). R, right hemisphere; L, left hemisphere.

Regarding the target detection task during the fMRI experiment, participants detected on average 77% of the 4 catch trials per session correctly. Accidental key presses during the presentation of the main stimuli were rare (2.5%).

fMRI Data

To identify the overall neuronal network involved in somatosensory perception, all tactile stimulation conditions were contrasted with the null events. In line with previous findings (Van Boven et al. 2005; Pleger et al. 2006; Beauchamp et al. 2007), this contrast revealed increased activation in contralateral SI in the postcentral gyrus (areas 3b, 1, 2) and in bilateral SII/parietal operculum (OP 1, 4), as well as in aIPS (areas hIP3, hIP2), inferior frontal gyrus (IFG; area 44), lateral prefrontal cortex (LPFC), pre-supplementary motor area (pre-SMA; area 6), insular cortex, thalamus, and cerebellum in both hemispheres (Figure 4.2 and Table 4.1).

Chapter 4. Feature-Specific Processing of Tactile Stimulus Attributes

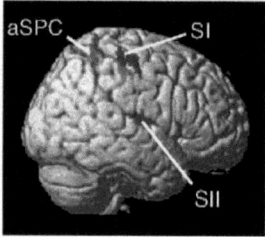
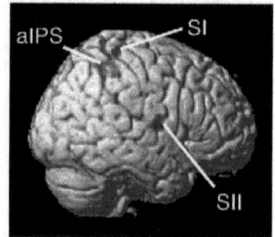

(a) Stimulus-specific differences   (b) Differential effects for motion direction and pattern orientation

Figure 4.3.: Differential effects within the network associated with tactile stimulation (shown in Figure 4.2). (a) Contrasting moving with stationary trials revealed increased activity in SI and SII (blue). Contrasting patterned with random stimuli showed increased activity in aSPC and SI (red). (b) Differential effects for motion direction of moving patterns were identified in SI and SII (blue). Differential effects for pattern orientation of stationary patterns were found in aIPS (red). SI, primary somatosensory cortex; SII, secondary somatosensory cortex; aSPC, anterior superior parietal cortex; aIPS, anterior intraparietal sulcus.

Assessing stimulus-specific differences within the network identified above, the comparison between moving and stationary stimuli revealed an increased BOLD response in contralateral SI (areas 3b, 1; $x = 46$, $y = -30$, $z = 60$) and SII (OP 4; $x = 48$, $y = -8$, $z = 10$), shown in Figure 4.3a (blue). Contrasting patterned with random stimulus trials revealed an increased activation in contralateral SI (areas 3b, 1, 2; $x = 42$, $y = -30$, $z = 56$) and anterior superior parietal cortex (aSPC; area 7; $x = 30$, $y = -48$, $z = 54$), shown in Figure 4.3a (red). For proper evaluation of the hemodynamic response profiles, BOLD time courses can be found in Appendix A (Figures A.1 and A.2). The analysis of interaction effects revealed no significant results.

We further investigated BOLD signal changes for the two possible pattern orientations (upward and downward diagonals). Within the activations of moving vs. stationary stimulation, the analysis revealed directionality differences for moving patterns in contralateral SI (area 1; $x = 38$, $y = -38$, $z = 64$) and SII (OP 4; $x = 52$, $y = -8$, $z = 8$), as shown in Figure 4.3b (blue). Within the activation map of patterned vs. random stimuli, differential orientation effects for stationary patterns were found in the right aIPS (area hIP3; $x = 36$, $y = -48$, $z = 50$), shown in Figure 4.3b (red). Closer inspection of these motion directionality and pattern orientation effects revealed that the upward diagonal motion direction showed by tendency increased activity in SI and SII compared with the downward diagonal motion direction. Likewise, the upward diagonal pattern orientation tended to be stronger represented in aIPS, compared with the downward diagonal pattern orientation. In terms of behavior (assessed outside the scanner), however, there were no significant differences between participants' performance in identifying upward and downward oriented patterns and motion directions.

Next, we tested differential effects of motion and pattern conditions in a whole-brain analysis, that is, also including areas that may not have been generally activated by tactile stimulation. The comparison between moving and stationary stimuli revealed increased activity in area hMT+/V5 ($x = -44$, $y = -70$, $z = -2$) and medial aSPC (area 5; $x = -10$, $y = -40$, $z = 56$) in the left hemisphere (Figure 4.4a, left). The hMT+/V5 activation remained also statistically significant after small volume correction with the anatomical map of hMT+/V5 defined using the Anatomy toolbox for SPM8. 15 % of this hMT+/V5 map were activated during tactile motion processing. The overlap is shown in Appendix B (Figure B.1). To assess to what extent these areas may have contributed to conscious processing of tactile motion, we investigated performance-dependent covariation in these areas. The BOLD responses in area hMT+/V5, but not medial aSPC, correlated positively with subjects' accuracy in identifying moving stimuli correctly (Figure 4.4a, right). Comparing patterned and random stimulus trials revealed increased activity in left IPC ($x = -60$, $y = -56$, $z = 32$), involving parts of the supramarginal and angular gyri (Figure 4.4b, left). BOLD time courses of the responses in both hMT+/V5 and IPC can be found in Appendix A (Figure A.3). Again, the BOLD responses in this area correlated positively with subjects' accuracy in identifying patterned stimuli correctly (Figure 4.4b, right). Additional bootstrapping analyses revealed that both correlations were significant (see Appendix C). Control analyses showed no correlation between motion-related responses in hMT+/V5 and participants' accuracy in pattern identification and between pattern-related responses in IPC and their accuracy in motion identification (both r's < 0.2, p's > 0.5). In addition, there were no other areas significantly activated in all presented contrasts besides the reported ones.

On the basis of the GLM results above, we studied possible changes in effective connectivity of the areas that were identified as specifically related to tactile motion and pattern processing. To this end, PPI analyses were performed using the peak voxels (spheres of 5 mm) of the motion and pattern activations located in left hMT+/V5 and left IPC, respectively, as seed regions for exploring coupling to the rest of the brain. The

Chapter 4. Feature-Specific Processing of Tactile Stimulus Attributes

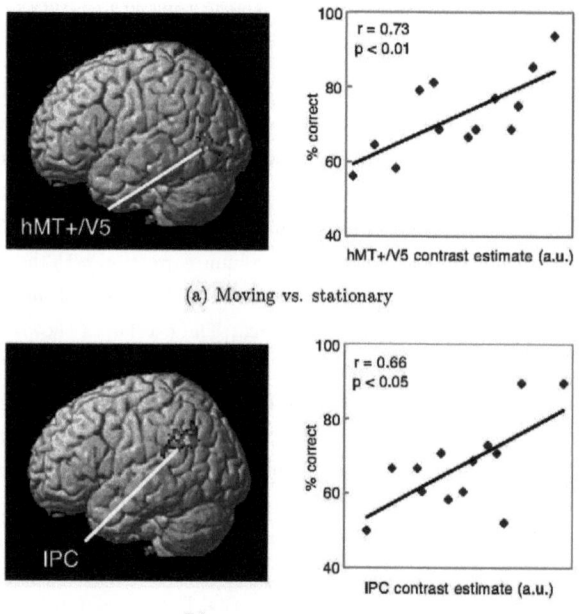

(a) Moving vs. stationary

(b) Patterned vs. random

Figure 4.4.: Areas involved in tactile motion and pattern processing. (a) Contrasting moving with stationary trials revealed increased BOLD responses in medial aSPC (not shown) and hMT+/V5. Individual subjects' contrast estimates in hMT+/V5 correlated positively with their accuracy in identifying moving stimuli correctly. (b) Contrasting patterned with random stimulus trials revealed increased BOLD responses in IPC. Individual subjects' contrast estimates in IPC correlated positively with their accuracy in identifying patterned stimuli correctly. hMT+/V5, middle temporal complex; IPC, inferior parietal cortex.

psychophysiological interaction term created for moving vs. stationary stimuli revealed a significant increase in coupling between left hMT+/V5, bilateral SI (ipsilateral: area 2; x = -48, y = -24, z = 42; contralateral: area 1; x = 56, y = -32, z = 52), and right aIPS (area hIP3; x = 30, y = -62, z = 46) during motion processing (Figure 4.5a). The PPI for patterned vs. random stimulation revealed a significant increase in effective connectivity between left IPC and right SI (areas 3b, 1, 2; x = 58, y = -22, z = 46) during pattern processing (Figure 4.5b).

4.4. Discussion

The first of the two studies providing the basis for the present thesis examined the neuronal networks underlying feature-specific processing of tactile stimulus attributes, namely motion and pattern, using high-field fMRI under tightly controlled passive stimulation conditions. Compared to matched control stimuli, stimulus-specific BOLD responses for both moving and patterned stimuli were already evident in early stages of somatosensory processing. In line with previous work, tactile motion evoked activity in hMT+/V5, an area traditionally associated with visual motion perception. An analysis of effective connectivity further revealed that the responses in area hMT+/V5 were functionally coupled to a somatosensory network including SI and aIPS (area hIP3). The processing of tactile patterns was characterized by distinct responses in IPC, and this activity was found to be functionally coupled directly to the responses in SI. Furthermore, both hMT+/V5 and IPC showed a significant correlation between task-induced neuronal activity and individual performance in identifying the respective stimulus attribute.

In general, tactile stimulation of the finger tip activated a widely distributed neuronal network including contralateral SI (areas 3b, 1, 2) and bilateral SII (OP 1, 4), aIPS (areas hIP3, hIP2), IFG (area 44), LPFC, as well as pre-SMA (area 6), insular cortex, thalamus, and the cerebellum, which altogether is in line with previously reported functional networks active during tactile processing (e.g., Stoesz et al. 2003; Van Boven et al. 2005; Blankenburg et al. 2006; Pleger et al. 2006; Beauchamp et al. 2007).

Within the overall network associated with tactile stimulation, we showed that processing of tactile motion was reflected in increased activity in contralateral SI (areas 3b, 1) and SII (OP 4). Moreover, we found both SI (area 1) and SII (OP 4) to respond differentially to different motion directions. Selective encoding of tactile motion by populations of SI neurons has been observed in invasive recordings in monkeys (Costanzo and Gardner

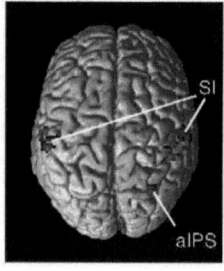 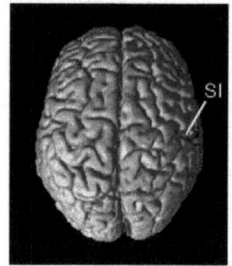

(a) Moving vs. stationary   (b) Patterned vs. random

Figure 4.5.: PPI analyses using left hMT+/V5 (a) and left IPC (b) as seed regions. (a) The interaction term for moving vs. stationary trials revealed a significant increase in coupling between left hMT+/V5, bilateral SI, and right aIPS during motion processing. (b) For patterned vs. random stimulus trials, the coupling between left IPC and right SI was significantly increased. SI, primary somatosensory cortex; aIPS, anterior intraparietal sulcus.

1980; Warren et al. 1986; Ruiz et al. 1995). More recently, Pei and colleagues (Pei et al. 2010) proposed that direction tuning in SI first emerges in area 3b, and is elaborated in area 1 to yield a more invariant representation of motion direction. In addition to SI, stimulus-specific firing of SII neurons during passive touch of moving gratings was shown in monkeys by Pruett and colleagues (Pruett et al. 2000). Whereas a general role of somatosensory cortex during tactile motion processing has also been reported in previous functional imaging work (Burton et al. 1999; Bodegård et al. 2000, 2001), the present evidence for differential BOLD signal changes for motion direction may not have been expected *a priori*. Thus far, direction- or orientation-selective responses in human sensory cortices have only been demonstrated using pattern classification methods for fMRI data analysis (Haynes and Rees 2005; Kamitani and Tong 2005, 2006). Because we did not investigate orientation encoding systematically in the present study and cannot fully exclude stimulus confounds, such as different skin indentation due to varying stimulus orientations, further studies are needed to explore the cause of the observed effect in more detail. However, the present evidence for motion-specific responses in human SI and SII complements the invasive findings and suggests that somatosensory cortex downstream from area 3b may similarly contribute to the emergence of an invariant representation of tactile motion in both species.

4.4. Discussion

The present analysis further revealed robust activation of area hMT+/V5 during processing of tactile motion. Area hMT+/V5 was long treated as a purely visual motion-sensitive area but has recently been shown to be engaged during processing of motion in other sensory modalities as well (Hagen et al. 2002; Blake et al. 2004; Summers et al. 2009). In Blake et al.'s work (Blake et al. 2004), a matched visual imagery condition did not activate hMT+/V5 significantly, which has been taken as evidence that the area's engagement during tactile motion cannot be fully explained by covert mental visualization of the somatosensory input. Involvement of area hMT+/V5 in tactile motion processing was consistently observed both in congenitally blind and in sighted subjects (Ricciardi et al. 2007; Matteau et al. 2010; Sani et al. 2010). Sani and colleagues further proposed a functional segregation of area hMT+/V5 in anterior and posterior subregions differentially involved in multisensory (visual and tactile) and visual motion processing (Sani et al. 2010), which might conform to areas MST and MT (Beauchamp et al. 2007). In line with this previous evidence, we found activation of hMT+/V5 extending towards anterior regions during processing of abstract, non-naturalistic Braille-like patterns, which may have been relatively difficult to imagine visually (see Deshpande et al. 2010 and Lacey et al. 2010 for related evidence). Furthermore, tactile motion in the present experiment was manipulated within a subset of non-target stimuli, while the subjects' task consisted of detecting a markedly distinct target pattern. This ensured that task demands did not encourage active visualization of the non-target's specific features. Notably, however, the hMT+/V5 responses evoked by moving stimuli covaried with subjects' ability to identify this type of stimuli under full attention outside the scanner. This performance-dependent covariation supports the view that hMT+/V5 may in particular contribute to conscious perception of tactile motion.

Our PPI analysis of effective connectivity revealed that activity in area hMT+/V5 was not only increased but also functionally coupled to the responses in SI (areas 1, 2) and aIPS (area hIP3) during tactile motion processing. The ventral part of the aIPS has been shown to be engaged in multisensory motion processing (Bremmer et al. 2001; Hagen et al. 2002; Summers et al. 2009), indicating that this area might be the human equivalent of the motion-sensitive area VIP within the monkey IPS. The present evidence for increased functional coupling of both primary somatosensory and motion-sensitive areas with hMT+/V5 may thus suggest direct transfer of somatosensory information to the motion-specialized area hMT+/V5 in visual cortex. Direct transfer in this context involves independence of the recruitment of other visual association areas such as the precuneus (Kosslyn et al. 1997; Goebel et al. 1998), which renders purely visual imagery as an explanation unlikely. Instead, the course of information processing might directly involve somatosensory cortex, aIPS, and hMT+/V5, given that neurons in monkey VIP receive projections from several visual areas (especially MT+/V5), and from motor, somatosensory, auditory, and other multisensory cortices (Lewis and Van Essen 2000). The present findings might indicate a more general role for hMT+/V5 in terms of multisensory motion processing (see also Sani et al. 2010). Similar conclusions have been drawn for visual and tactile object processing in the LOC (Deshpande et al. 2008; Lacey and Sathian 2011).

In addition to the motion-specific responses in SI, SII, and in hMT+/V5, moving stimuli evoked increased activity in the medial part of the left aSPC (area 5). The superior parietal cortex is known to be involved in various cognitive processes, in particular somatosensory and sensorimotor integration as well as visuospatial attention and memory. The aSPC was shown to integrate information mainly from the somatosensory cortex (Pandya and Seltzer 1982; Scheperjans et al. 2005). In the present experimental context, medial aSPC activity neither covaried with subjects' behavioral performance nor reflected the stimuli's motion direction, and was not functionally coupled to the motion network outlined above. This activation thus likely reflected a less direct involvement in somatosensory information processing.

Like tactile motion, tactile patterns evoked increased activity in SI (areas 3b, 1, 2), compared to randomly structured control stimuli, which is in agreement with previous work on tactile form perception in humans (Bodegård et al. 2001; Van Boven et al. 2005) and monkeys (Phillips et al. 1988; Hsiao et al. 1993; DiCarlo and Johnson 2002). Pattern processing further engaged the aSPC (area 7), corroborating previous evidence that this part of the aSPC, which receives direct projections from area 2 in SI (Pandya and Seltzer 1982; Scheperjans et al. 2005), is critically involved in somatosensory processing of shape information (Stoeckel et al. 2003). Complementing electrophysiological evidence for the existence of orientation-tuned neurons in somatosensory areas (DiCarlo and Johnson 2000; Hsiao et al. 2002; Fitzgerald et al. 2006; Bensmaia et al. 2008), we found differential BOLD responses to different pattern orientations in aIPS (area hIP3), which underpins this multisensory area's role in the processing of tactile object features. As with the differential effects for motion direction, the present finding of differences for pattern orientations using fMRI was rather unexpected. There is, however, recent evidence for coarse-scale orientation maps in V1 measured using fMRI (Freeman et al. 2011), which indicates that

4.4. Discussion

selective population responses may in principle be assessed also with univariate statistical analysis. Although their functional significance has to be investigated in further studies, these findings support the existence of a putative human equivalent of the monkey AIP, which is known to be engaged in tactile and visual object processing (Grefkes and Fink 2005).

In addition to the pattern-specific responses in SI, aSPC, and aIPS, which were also activated during tactile stimulation per se, our whole-brain analysis revealed that tactile pattern processing further recruited an area in the left IPC, including parts of the supramarginal and angular gyri. Associative somatosensory function was attributed to the IPC in several lesion studies (Reed and Caselli 1994; Reed et al. 1996; Nakamura et al. 1998). For instance, Reed and colleagues (Reed et al. 1996) reported a patient's impairment of shape recognition specific to the tactile modality resulting from a small inferior parietal infarction. In an fMRI study by Deibert and colleagues (Deibert et al. 1999), the inferior parietal lobule (supramarginal and angular gyrus) was activated when subjects manipulated and identified an object's shape. More recently, Miquée and colleagues (Miquée et al. 2008) investigated fMRI correlates of haptic shape perception in a task divided in shape-encoding and -matching steps, and showed that the IPC was specifically involved in shape matching. The latter finding supports the view of the IPC as a neuronal substrate of shape representation rather than coordination of finger movements as required in tactile object recognition tasks. The present results confirm this conclusion, demonstrating IPC activity during passive touch of patterned Braille-like stimuli. Furthermore, the BOLD responses in IPC covaried with participants' individual ability to identify patterned stimuli (assessed outside the scanner), providing evidence that the IPC may in particular contribute to conscious perception of tactile patterns. Finally, our PPI analysis showed that pattern processing entailed specific functional interactions between IPC and contralateral SI, which indicates that the IPC was intimately involved in the exchange of modality-specific somatosensory information. This exchange of tactile information might potentially occur by projecting from SI to IPC via SII, given the evidence for anatomical connections between SI and SII as well as between SII and IPC (Eickhoff et al. 2010). Using a less stringent significance threshold of $p<0.005$ (uncorrected) for our PPI analysis, we found suggestive evidence for a role of SII as part of this specific functional network supporting this hypothesis, which remains to be explored in the future.

The engagement of IPC during tactile pattern processing was in many respects phenomenologically similar to the engagement of hMT+/V5 during tactile motion process-

Chapter 4. Feature-Specific Processing of Tactile Stimulus Attributes

ing. Both areas were selectively recruited by the presence of an abstract tactile stimulus attribute (pattern or motion), were functionally coupled to somatosensory cortex in a stimulus-dependent manner, and the specific responses in both areas covaried with subjects' perceptual performance. With respect to the particular significance of these areas in the processing of complex stimulus attributes, the present results may suggest that in analogy to the visual system, modality-specific somatosensory areas may interact with regions that are dedicated to the integration of specific perceptual features, such as motion or pattern, into a conscious perceptual concept. Such a concept might not necessarily be modality-specific and seems to involve not only designated somatosensory (IPC) or multisensory areas (aIPS) but, in the case of tactile motion, also area hMT+/V5.

In sum, our results corroborate that somesthesis of complex stimulus attributes engages characteristic processing networks that, on the one hand, involve modality-specific somatosensory areas but, on the other hand, incorporate multisensory or even acknowledged visual areas. This overall picture is in line with increasing evidence that processing of complex sensory input from any specific modality may result in an abstract, essentially multisensory representation (e.g., Amedi et al. 2001; Grefkes et al. 2002; Hagen et al. 2002; Beauchamp et al. 2007; Deshpande et al. 2008; Summers et al. 2009; Lucan et al. 2010; Lacey and Sathian 2011). Thereby, areas specialized in the processing of abstract object attributes such as motion and pattern may be more process- than modality-driven. The present findings support this integrative view, and indicate that early, modality-specific representations of complex tactile information may be directly relayed to cortical areas that are dedicated to the further processing of specific stimulus features, regardless of their sensory modality.

Chapter 5.

Top-Down Attentional Bias for Gating Tactile Perception

As described in detail in Section 2.5, selective attention is the cognitive process of directing mental efforts on specific aspects of the world while ignoring other things. Attention involves the allocation of processing resources and can be controlled by cognitive factors such as expectation and current goals. Attentional mechanisms provide us with the ability to selectively suppress irrelevant information and strengthen the representations of those aspects that are currently relevant to our intentions.

In this chapter, we present the second experimental investigation, which focusses on the cortical mechanisms of selective attention involving top-down directed control of somatosensory processing for the perception of tactile stimulus attributes. In a short introduction, we review the current knowledge in the field and give a motivation for this investigation. Following a detailed explanation of the methods used, we present the results of this study. Finally, conclusions are drawn including a discussion of their significance and how they relate to the current knowledge in the field.

5.1. Introduction

In the human brain, a network of brain areas distributed across frontal and parietal cortices subserves cognitive control of action and attention. This frontoparietal network involves inferior and superior parietal areas as well as the prefrontal cortex. It commonly activates during many kinds of cognitive demands, which indicates dependence on similar control mechanisms (reviewed in Duncan and Owen 2000 or Duncan 2010). An important quality of this global network is its flexible adaptability, which enables us to focus our

Chapter 5. Top-Down Attentional Bias for Gating Tactile Perception

attention on currently relevant information. The representation of task-relevant information seems to be held by several parts of the frontoparietal network, which are highly adaptable. Neurophysiological recordings in monkeys showed that neurons in the LPFC reflected the currently relevant category of a visual stimulus, while becoming unresponsive to this category when the task demands changed (Freedman et al. 2001). Functional imaging work in humans complements these findings by identifying specific responses to task-relevant information in a distributed network of frontal and parietal areas (Hon et al. 2006; Hampshire et al. 2007; Thompson and Duncan 2009). Despite these recent advances in delineating frontoparietal function, the specific contributions made by the individual areas comprising the frontoparietal network remain to be elucidated.

Promising work in this context suggests that the IFG in the human LPFC might play a prominent role in this network. Several functional imaging studies show category- or target-selective activity in this region (Jiang et al. 2007; Hampshire et al. 2009), which further generalizes to the role of the IFG during response inhibition (Hampshire et al. 2010). Hampshire and colleagues revealed that the IFG is not specifically involved in inhibitory control but more generally in the detection of task-relevant cues. In addition, there is evidence supporting the view that this region may be able to mediate cognitive control over other brain areas (Staines et al. 2002; Li et al. 2007; Duann et al. 2009). The IFG appears therefore ideally suited to act as the neural substrate for the coordination of bottom-up, stimulus-driven processes and the processing of top-down, goal-directed intentions (for related evidence see Asplund et al. 2010). However, the specific way, in which this coordination may occur remains poorly understood.

The current knowledge of frontoparietal function mainly derives from investigations employing visual task conditions. This raises the question whether similar principles can also be transferred to the somatosensory modality. Attention in general has been shown to enhance activity across modalities specifically in those areas that are normally driven by the kind of stimulation, which is attended to (Corbetta et al. 1991; Desimone and Duncan 1995; Burton et al. 1999). In anticipation of sensory input attention serves to increase the processing capacities of specific neural populations (Desimone and Duncan 1995). Thus, in the tactile domain attentional effects can be expected in SI and SII, as demonstrated by several studies using fMRI in humans (Mima et al. 1998; Burton et al. 1999; Johansen-Berg et al. 2000), including the one presented in Chapter 4. Increased activity due to attended tactile stimulus attributes was further observed in inferior and superior parietal areas as well as in prefrontal cortex (Van Boven et al. 2005; Burton et al. 2008), which may suggest that sensory areas interact with the frontoparietal network for attentional task control. Burton and colleagues reasoned that activity in frontal and parietal areas indicates resource mobilization for target or response selection (Burton et al. 2008). These cognitive components are involved whenever tactile or visual stimulation attributes are discriminated. However, the causality of the activity in areas involved in tactile task performance is not yet clear, as well as possible interactions between frontoparietal cortex and sensory areas that create the prerequisites for task accomplishment.

Here, we investigated the cortical network involved in a tactile change-detection task that required selective attention to either felt spatial pattern or felt temporal frequency. Tactile stimulation was presented via a finger-tip-sized multi-pin stimulation device similar to a Braille display. This allowed the administration of complex tactile stimuli to the left index finger tip, comprising one or other of two specific spatial patterns that could be presented at one or other of two different temporal frequencies. The task was to attend to either the spatial pattern or the temporal frequency for a successive train of stimuli, in order to detect any occasional change in that respective stimulus attribute, while ignoring any change in the currently irrelevant attribute. We assessed fMRI activations as a function of task (spatial pattern or temporal frequency attended), and also as a function of change-detection for the attended stimulus attribute. The amplification of similar tactile stimulation in all conditions allowed investigating task- or detection-specific activations independent of sensory input. Tactile stimulation and response execution were decoupled in order to allow for separation of activations due to the somatosensory stimuli and tasks, versus due to key presses with the right hand.

Based on previous research, we expected not only somatosensory but also prefrontal and parietal cortex to be involved in tactile task performance. We further anticipated that frontoparietal components of this task-related network might selectively respond to the detection of change in the task-relevant stimulus attribute. In addition, we studied any task-dependent changes in the effective connectivity between the brain areas implicated. The analysis aimed to characterize the network interactions involved in top-down control of somatosensory processing as assessed in terms of PPIs (Friston et al. 1997). Furthermore, we performed more specific network analyses to elucidate in particular the causal directional influences that the prefrontal cortex may exert over somatosensory areas for such top-down modulation using DCM (Friston et al. 2003) and BMS (Penny et al. 2004; Stephan et al. 2009).

5.2. Materials and Methods

Participants

Eleven healthy volunteers (aged between 24 and 32 years; nine females, one left-handed) participated in the study with written informed consent. The study corresponded to the Human Subjects Guidelines of the Declaration of Helsinki and was approved by the Ethical Committee at the faculty of medicine, Charité University Hospital Berlin.

Stimuli & Tasks

Tactile stimulation was applied to the left index finger by the 16-dot piezoelectric Braille-like display described in Section 4.2. On each trial, the pins of the display were driven for 1 s by a 160 Hz sinusoidal carrier signal, which was amplitude-modulated by a sine function of either 28 or 40 Hz to yield two different temporal frequencies. Two different maximal amplitudes of pin-movement (1.1 and 0.9 mm) could be assigned to each pin of the display in such a way that it formed one of two possible spatial patterns (as shown schematically in Figure 5.1a). After half of the presentation time, both stimulus attributes (pattern and frequency) could change independently from each other, e.g., the frequency attribute changed from 40 to 28 Hz and/or the pattern attribute changed from the one to the other.

Participants performed three different tasks. Two of these tasks involved selective attention to the stimulus attributes of pattern (P task) or frequency (F task). In both tasks, participants judged whether a change in the cued feature had occured or not, by pressing one of two response keys with the right index or middle finger. Note that change or lack of change in the two stimulus attributes occurred independently of each other, regardless of the currently performed task. Finally, in the third control task, participants also received tactile stimulation, but were now asked to pay no attention to it and to press one of the two response keys arbitrarily instead when prompted to respond after each stimulus. Stimulus presentation and task cueing was controlled using custom MATLAB code (v7.5, The MathWorks, Natick, MA) and the Cogent 2000 toolbox (http://www.vislab.ucl.ac.uk/cogent.php).

Design & Procedure

Prior to the experiment, participants were familiarized with the stimulation device and with the different tasks. The actual fMRI experiment consisted of four sessions. Each session comprised ten blocks of eight trials each. In each block, participants performed one

5.2. Materials and Methods

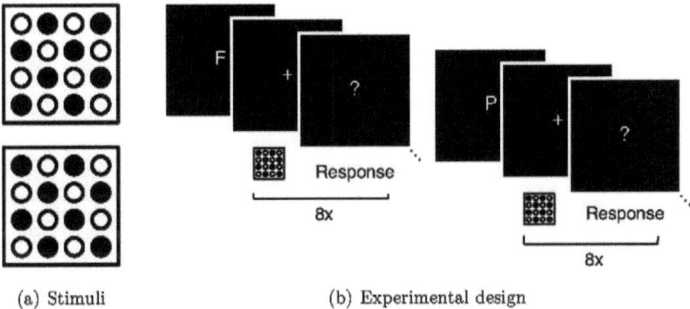

(a) Stimuli   (b) Experimental design

Figure 5.1.: Stimuli and experimental design. (a) The pins of the tactile display were driven by a sinusoidal carrier signal, which was amplitude-modulated by a sine function of either 28 or 40 Hz. Two different maximal amplitudes (black and white pins) were assigned to the display's pins in such a way that they formed one of two possible patterns. (b) One session comprised ten blocks of 8 trials. In each block, participants performed one of the three tasks cued at the beginning of the block ('P', 'F', or 'O'). The onset of the motor response was cued visually ('?') at a randomly chosen time point within the ISI in order to dissociate responses and tactile stimulation.

of the three tasks as instructed by a symbolic visual letter cue at the beginning of the block ('P' = P task, 'F' = F task, 'O' = control task; see Figure 5.1b). The blocks were arranged in pseudo-random serial order, such that one session comprised four blocks of the P task, four blocks of the F task, and two blocks of the control task. All trials had a duration of 1 s and were presented in pseudo-random serial order, such that each of the 16 possible combinations[1] of feature changes appeared once in two attention blocks (P or F task). The control blocks included only trials without feature changes to discourage participants from inadvertently or implicitly performing one of the two attention tasks during control blocks. The ISI was randomly varied between 5 and 13 s. In order to dissociate tactile stimulation and motor responses, participants did not respond immediately after tactile stimulation, but the onset of the key press was cued visually ('?') at a randomly chosen time point (between 2 and 8 s after stimulation offset) within the ISI (Figure 5.1b).

---

[1] One of two possible patterns and one of two possible frequencies for each half of the stimulation come to a total of (2x2) x (2x2) = 16 combinations.

Chapter 5. Top-Down Attentional Bias for Gating Tactile Perception

fMRI Data Acquisition

Functional imaging was performed on a 1.5 Tesla Siemens Sonata MRI scanner (Siemens Medical Solutions, Erlangen, Germany) with a standard head-coil system. T_2^*-weighted functional images were acquired using an EPI sequence (TR = 2010 ms, TE = 40 ms, flip angle = 90°). For each session, 430 EPI volumes were obtained, each consisting of 36 axial slices covering the whole brain in an interleaved slice ordering (voxel size 3x3x3 mm, matrix size 64x64). Visual cues were presented on a screen, visible from the scanner via a mirror system attached to the head coil.

Data Preprocessing

Functional images were preprocessed and analyzed using SPM8 (Wellcome Department of Imaging Neuroscience, University College London, UK). The first four volumes of each experimental session were discarded in order to allow the MR signal to reach equilibrium. To minimize movement-induced image distortions, each data set was realigned to the first image of the first session as described in Section 3.3. The realigned images were spatially normalized to the standard MNI template brain and smoothed using an isotropic, three-dimensional Gaussian kernel of 7 mm FWHM, in accord with the standard SPM approach. To remove global effects from the fMRI time series, detrending was applied based on LMGS (Macey et al. 2004). The images were high-pass filtered (cut-off frequency 1/128 Hz) in order to remove low-frequency signal drifts. To reduce high-frequency noise, serial correlations were modeled using an autoregressive model.

Statistical Data Analysis

A standard two-level mixed-effects model (Friston et al. 1995) as introduced in Section 3.3 was used for statistical analysis. At the first level, multiple regression within the framework of the GLM was used to implement a within-subject analysis. For each data set BOLD responses were modeled by nine stick functions indicating the onsets of the tactile stimuli, i.e., four types of correctly judged P task trials, four types of correctly judged F task trials, and one stick function for control task trials. The four types of P and F task trials were (1) stimuli with change in pattern and frequency, (2) stimuli with change in pattern only, (3) stimuli with change in frequency only, and (4) stimuli without change in any feature. Three additional stimulus functions for nuisance effects comprised incorrect trials, motor responses, and visual task instructions. These regressors were then convolved with a standard HRF and included in the GLM. To account for occasional signal intensity changes within slices due to susceptibility-induced frequency variations, nine additional nuisance regressors were included. These corresponded to the first nine eigenvariates, which were extracted exclusively from signals outside the brain in a previous SPM analysis (thresholded at $p<0.05$). After fitting the model to the experimental data, contrast images were generated from the stimulus functions' parameter estimates for each of the stimulus types. At the second level, the individual subjects' contrast images were entered into a 2x4 within-subjects ANOVA with factors attention (P task or F task) and change (change in both features, change in pattern, change in frequency, or no change). This allowed computing differential effects between the different stimulus types of both attention tasks using contrast vectors to produce SPMs. To investigate main effects of and differential effects between attention (P and F) tasks and control task, another second-level model was implemented using a paired t-test.

Conjunction Analysis

To assess the effects of detecting change in the attended stimulus attribute while holding sensory input constant, a conjunction analysis of contrasts was implemented (Price and Friston 1997). To this end, the same mixed-effects model was rerun with a single modification at the first level concerning the two stimulus functions for stimuli with change in both features. For both regressors, the trials were divided equally and arbitrarily into two distinct new stimulus functions. This separation allowed for independent computation of the conjunction contrasts, which required separate trials as baselines. At the second level, this modification now resulted in a 2x5 within-subjects ANOVA. This allowed computing the contrasts of (1) P task trials with change in pattern vs. F task trials with change in pattern and (2) F task trials with change in frequency vs. P task trials with change in frequency. A conjunction analysis of (1) and (2) now revealed areas involved in detecting the change in a specific stimulus attribute independent of sensory input and type of attended attribute.

PPI Analysis

To investigate the functional coupling of areas involved in selective attention, we performed a PPI analysis (Friston et al. 1997) as introduced in Section 3.4. Spheres with a radius of 5 mm constructed around peak voxels of the individual subjects' activations located in right SI ($x = 48$, $y = -34$, $z = 52$) served as seed regions for extracting the first eigenvariate of the signal. At the subject level, the physiological variable was extracted and the psychophysiological interaction term was created for attention (P and F task) vs. no attention (control task). Subsequently, this term was entered into a GLM. At the group level, contrast images of the PPIs of the individual subjects were analyzed using one-sample t-tests.

Chapter 5. Top-Down Attentional Bias for Gating Tactile Perception

Statistical Inference

All reported coordinates correspond to the anatomical MNI space. The SPM anatomy toolbox (Eickhoff et al. 2005) was used to establish cytoarchitectonic reference. To investigate main effects of tasks and differential effects between tasks and trials, we used a significance threshold of $p_{cluster} < 0.05$, corrected for multiple comparisons. For the conjunction analysis, we adjusted our significance threshold to $p < 0.001$, uncorrected, as we were testing a specific hypothesis rather than searching the whole brain. For the PPI analysis, the activation pattern revealed by contrasting all tactile stimulation trials with baseline was used as a mask (mask $p_{cluster} < 0.05$, corrected for multiple comparisons), and the significance threshold of $p < 0.001$, uncorrected was chosen for statistical inference.

Dynamic Causal Modeling

The effective connectivity between the lateral prefrontal cortex and somatosensory areas was assessed using DCM (Friston et al. 2003; Stephan and Friston 2010). Functional imaging data were remodeled using a GLM with three regressors of interest, which encoded (1) all tactile stimulation trials, (2) all attention task trials, and (3) all trials with change in the attended stimulus attribute. Three additional regressors for nuisance effects comprised incorrect trials, motor responses, and visual task instructions. Spheres with a radius of 7 mm constructed around subject-specific peaks of the activations located in right SI (mean: $x = 52$, $y = -30$, $z = 54$), right SII (mean: $x = 62$, $y = -16$, $z = 20$), and right IFG (mean: $x = 54$, $y = 10$, $z = 20$) served as seed regions for extracting the first eigenvariate of the signal. These time series were fed into separate models for each subject. The intrinsic coupling of the network included reciprocal connections between SI and SII as well as between SII and IFG (as in Figure 5.6a), based on known interconnectivity of these areas (Eickhoff et al. 2010). Three conditions described the influence on effective connectivity in this network: (1) tactile stimulation, which entered the system as driving input via SI, (2) active attention (to either pattern or frequency), which could act as driving input to IFG and/or as modulatory influence on effective connectivity, and (3) change in the attended stimulus attribute as modulatory input. Ten different models were generated using different combinations of driving and modulatory input (see Table 5.3). To identify the model explaining the data best, BMS was performed using a random-effects analysis, which computed the exceedance probability that one model is more likely than any other model, given the group data (Stephan et al. 2009, 2010). For group-level inference on parameter estimates Bayesian parameter averaging (BPA) was used, which generated the joint posterior density of a given model's parameter estimates for the entire

5.3. Results

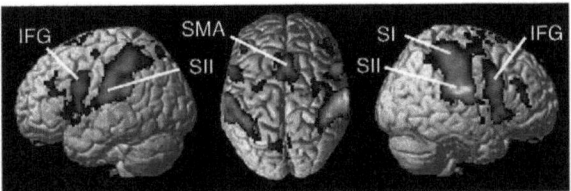

(a) Attention (P and F task) vs. baseline

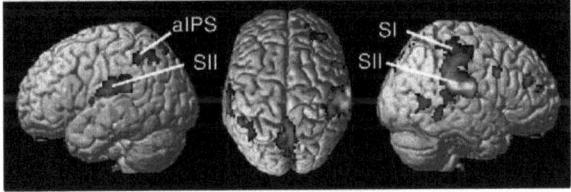

(b) Control vs. baseline

Figure 5.2.: Overall neuronal networks involved in the different tasks. (a) Contrasting all tactile stimulation trials of both attention tasks with baseline revealed a distributed network activated during task-relevant stimulation. (b) Contrasting all tactile stimulation trials of the control task with baseline revealed a reduced network activated during passive stimulation. SI, primary somatosensory cortex; SII, secondary somatosensory cortex; aIPS, anterior intraparietal sulcus; IFG, inferior frontal gyrus; SMA, supplementary motor area.

group by combining the individual posterior densities (Stephan et al. 2010). A significance threshold of $p<0.05$ was used for the connectivity analysis.

5.3. Results

Behavioral Data
On average, 71% (SEM = ±3%) and 66% (SEM = ±2%) of the stimuli were correctly classified in the P and in the F task, respectively. Performance was significantly higher than chance level (50%) in both attention tasks (both p's < 0.001, two-tailed t-test), and there was no significant difference between P and F task performance ($p > 0.1$). The average response time was 638 ms, 627 ms, and 688 ms after response cue onset for P, F, and control task, respectively. Again, there was no significant difference between P and F task ($p > 0.3$), but participants responded significantly faster in both attention tasks

Chapter 5. Top-Down Attentional Bias for Gating Tactile Perception

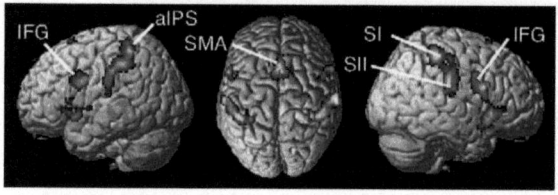
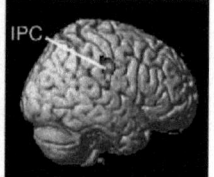

(a) Attention (P and F task) vs. control task   (b) P task vs. F task

Figure 5.3.: Differential effects between the tasks. (a) The comparison between both attention tasks and the control task revealed increased activation in SI and SII, as well as in aIPS, IFG, insular cortex, and SMA. (b) Contrasting P task with F task revealed a single activated cluster in IPC extending to SI. SI, primary somatosensory cortex; SII, secondary somatosensory cortex; aIPS, anterior intraparietal sulcus; IFG, inferior frontal gyrus; SMA, supplementary motor area.

than in the control task (both p's < 0.05), which might arise if they did not select which response to make until the prompt in the latter case.

fMRI Data

First, we identified the overall neuronal networks involved in the different tasks. Contrasting all tactile stimulation trials of both attention tasks with baseline revealed the expected increased activation in contralateral SI in the postcentral gyrus (areas 1, 2) as well as in SII/parietal operculum (OP 1, 4), aIPS (areas hIP2, hIP3), LPFC, IFG (area 44), supplementary motor area (SMA; area 6), and insular cortex in both hemispheres, and in contralateral thalamus (Figure 5.2a and Table 5.1). This network was thus implicated in the two tasks involving selective attention to a specific tactile stimulus attribute. Contrasting all tactile stimulation trials of the control task with baseline revealed increased activation in contralateral SI (areas 1, 2) and in bilateral SII (OP 1, 4), as well as small activation clusters in left aIPS (areas hIP1, hIP2), right LPFC, right IFG (area 44), and precuneus (Figure 5.2b and Table 5.1) during passive tactile stimulation. These regions were therefore implicated in a task that did not require attention to a specific tactile stimulus attribute. Note that these activations were well separated from the required motor responses as an investigation of the nuisance regressor comprising key press onsets revealed (Figure B.2 in Appendix B).

We further investigated differential effects between the tasks. The comparison between both active attention tasks and the control task revealed increased activation in contralat-

5.3. Results

Attention (P and F task) vs. baseline					
Region	Hemisphere	x	y	z	T-value
Primary somatosensory cortex	R	46	-34	52	10.51
Secondary somatosensory cortex	R	64	-16	20	31.87
	L	-60	-18	26	12.79
Anterior intraparietal sulcus	R	44	-44	58	15.24
	L	-46	-44	54	12.90
Lateral prefrontal cortex	R	54	34	26	7.57
	L	-44	34	20	6.40
Inferior frontal gyrus	R	52	12	30	14.90
	L	-60	10	20	10.69
Supplementary motor area	R/L	4	8	60	11.47
Insular cortex	R	38	20	0	9.75
	L	-32	20	-2	8.44
Thalamus	R	12	-20	10	8.85
Control vs. baseline					
Region	Hemisphere	x	y	z	T-value
Primary somatosensory cortex	R	48	-34	58	4.73
Secondary somatosensory cortex	R	66	-16	20	19.11
	L	-62	-12	18	5.87
Anterior intraparietal sulcus	L	-50	-50	50	8.75
Lateral prefrontal cortex	R	36	42	34	6.77
Inferior frontal gyrus	R	52	12	32	7.11
Precuneus	R/L	0	-50	42	7.98
Attention (P and F task) vs. control task					
Region	Hemisphere	x	y	z	T-value
Primary somatosensory cortex	R	46	-36	52	5.08
Secondary somatosensory cortex	R	64	-16	20	9.02
Anterior intraparietal sulcus	R	44	-44	58	7.41
	L	-46	-44	54	10.98
Inferior frontal gyrus	R	52	12	28	6.93
	L	-60	10	20	6.43
Supplementary motor area	R/L	6	10	60	7.46
Insular cortex	R	40	20	-2	8.14
	L	-32	20	-2	7.47
Thalamus	R	12	-14	0	6.20

Table 5.1.: Functional regions active during the different tasks. x, y, z are MNI coordinates (mm). T-values are local maxima within a significant cluster of activated voxels with $p_{cluster} < 0.05$, corrected for multiple comparisons (group-level analysis). R, right hemisphere; L, left hemisphere.

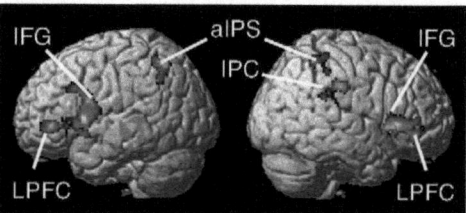
(a) Change vs. no change

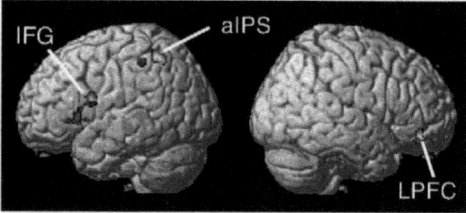
(b) Conjunction analysis

Figure 5.4.: Differential effects of change in the attended stimulus attribute. (a) Contrasting trials with change in the attended stimulus attribute and trials without change in the attended feature revealed a frontoparietal network of increased activity. (b) The conjunction analysis of (1) P task trials with change in pattern vs. F task trials with change in pattern and (2) F task trials with change in frequency vs. P task trials with change in frequency revealed a similar activation pattern. IPC, inferior parietal cortex; aIPS, anterior intraparietal sulcus; IFG, inferior frontal gyrus; LPFC, lateral prefrontal cortex.

5.3. Results

Change vs. no change					
Region	Hemisphere	x	y	z	T-value
Anterior intraparietal sulcus	R	50	-50	56	3.91
	L	-40	-52	54	4.07
Lateral prefrontal cortex	R	46	36	-6	4.35
	L	-50	40	-2	5.11
Inferior frontal gyrus	R	56	18	0	4.67
	L	-54	10	0	3.65
Insular cortex	R	32	22	0	4.45
	L	-34	26	2	3.70
Inferior parietal cortex	R	62	-24	34	4.39
Conjunction analysis					
Region	Hemisphere	x	y	z	T-value
Anterior intraparietal sulcus	L	-40	-50	56	2.89
Lateral prefrontal cortex	R	38	42	-12	2.65
Inferior frontal gyrus	L	-50	6	20	2.84
Insular cortex	L	-32	26	0	3.10

Table 5.2.: Functional regions active during change detection. x, y, z are MNI coordinates (mm). T-values are local maxima within a significant cluster of activated voxels with $p_{cluster} < 0.05$, corrected for multiple comparisons (upper table) and $p < 0.001$, uncorrected (lower table); group level analyses. R, right hemisphere; L, left hemisphere.

Chapter 5. Top-Down Attentional Bias for Gating Tactile Perception

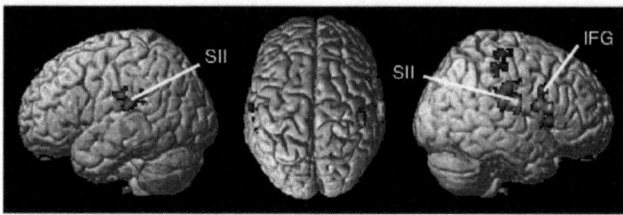

Figure 5.5.: PPI analysis for selective attention using right SI as seed region. The PPI term created for attention vs. control revealed a significant increase in coupling between right SI, bilateral SII, and right IFG during selective attention to tactile stimulus attributes. SII, secondary somatosensory cortex; IFG, inferior frontal gyrus.

eral SI (areas 1, 2) and SII (OP 1, 4), as well as in aIPS (areas hIP2, hIP3), IFG (area 44), insular cortex, and SMA (area 6) in both hemispheres, as well as in contralateral thalamus (Figure 5.3a and Table 5.1). This implicated these areas in selective attention to any feature of the tactile stimulation independent of sensory input. Contrasting P task trials with F task trials revealed one single activation cluster in right IPC (x = 62, y = -18, z = 24; Figure 5.3b), which extends to SI (areas 1, 2). The opposite comparison did not reveal any significant activation at our corrected threshold.

Next, we tested differential effects of change in the attended stimulus attribute. To this end, all trials with correctly judged change in the attended stimulus attribute were contrasted vs. trials without change in the attended feature, pooled across P and F task and regardless of change in the non-attended feature. This contrast revealed increased activation in a frontoparietal network including bilateral IFG (area 44), aIPS (areas hIP2, hIP3), LPFC, and insular cortex, as well as the right IPC (Figure 5.4a and Table 5.2) for detecting change in the attended stimulus attribute independent of change in the non-attended stimulus attribute. Moreover, a comparison between trials with change in the non-attended stimulus attribute and trials without change in the non-attended feature revealed no significant results. To further support the ensuing assumption that this effect was the result of top-down control of attention, a conjunction analysis of contrasts (Price and Friston 1997) was performed. For this purpose, the contrasts of (1) trials with correctly judged change in pattern for P task vs. trials with change in pattern for F task and (2) trials with correctly judged change in frequency for F task vs. trials with change in frequency for P task were computed. A conjunction analysis of (1) and (2) would reveal areas involved in detecting change in the attended stimulus attribute independent

5.3. Results

of sensory input and type of attended feature. Applying this kind of analysis we identified activation clusters in aIPS (areas hIP2, hIP3), IFG (area 44), and insular cortex in the left hemisphere, as well as in right LPFC (Figure 5.4b and Table 5.2). This activation pattern overlaps with the frontoparietal network identified for change detection, which is shown in Figure 5.4a.

In order to investigate inter-regional interplay during task performance, we next studied the effective connectivity of areas that were implicated in selective attention. To this end, we performed a PPI analysis using the peak voxels (spheres of 5 mm radius) of right SI as seed regions for exploring the coupling to areas generally involved in the task. The PPI term created for attention (P and F task trials) vs. no attention (control task trials) revealed a significant increase in coupling between right SI, bilateral SII (OP 1, 4; contralateral: $x = 60$, $y = -14$, $z = 22$; ipsilateral: $x = -64$, $y = -20$, $z = 26$), and right IFG (area 44; $x = 50$, $y = 8$, $z = 16$) during selective attention to either tactile stimulus attribute (Figure 5.5).

As revealed by the PPI analysis and anticipated based on previous evidence (e.g., Staines et al. 2002; Asplund et al. 2010; Hampshire et al. 2010), SI, SII, and IFG appeared to play key roles during task performance. To investigate the causal, directional influences of the experimental manipulations on effective connectivity between these areas, network models were created based on known interconnectivity (Eickhoff et al. 2010) of these areas for use in DCM (Figure 5.6a). Three conditions described the influence on effective connectivity: (1) tactile stimulation, which entered the system as driving input via SI, (2) active attention (to either pattern or frequency), which could act as driving input to IFG and/or as modulatory influence on effective connectivity, and (3) change in the attended stimulus attribute as modulatory input. For completeness, and to allow formal model comparison, ten different models were generated using different combinations of driving and modulatory input (Table 5.3). In order to identify the model explaining the data best, BMS was performed using a random-effects analysis (Figure 5.6b). The optimal model (shown in Figure 5.6a) clearly outperformed the others, as can be seen in Figure 5.6b. Selective attention to either stimulus attribute significantly increased the directional top-down influence from IFG to SII in this network. Change in the attended stimulus attribute significantly enhanced the strength of the feedforward connection from SII to IFG.

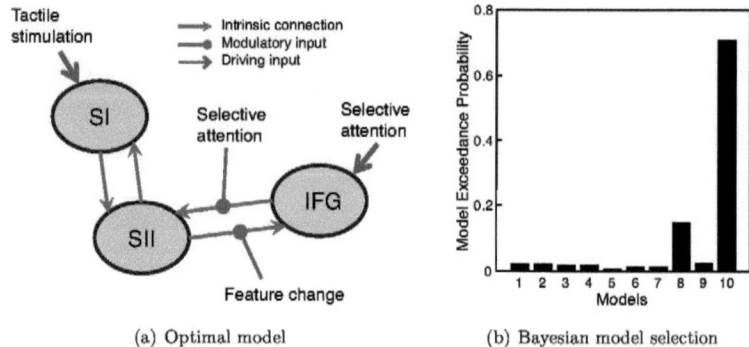

(a) Optimal model (b) Bayesian model selection

Figure 5.6.: Analysis of effective connectivity in a network consisting of SI, SII, and IFG. (a) The optimal model involves enhanced feedback connectivity from IFG to SII due to attention and increased feedforward connectivity from SII to IFG due to change in the attended feature. (b) BMS results are plotted by means of exceedance probabilities (the model shown in (a) is model 10). SI, primary somatosensory cortex; SII, secondary somatosensory cortex; IFG, inferior frontal gyrus.

5.4. Discussion

In the second study presented in this thesis we took advantage of fMRI to examine the cortical network associated with a tactile change-detection task that required selective attention to a specific tactile stimulus attribute. The participants' task was to attend either to the spatial pattern or to the temporal frequency of complex Braille-like tactile stimuli and to detect changes in the respective stimulus attribute. This task involved both the deployment of top-down directed selective attention and the identification of task-relevant bottom-up sensory information. Investigating task-dependent activations we found a distributed network of somatosensory as well as prefrontal and parietal areas underlying tactile task performance. In addition, frontoparietal components of this network including IFG, LPFC, and aIPS selectively responded to the detection of change in the task-relevant stimulus attribute. PPI analysis further revealed that SI, SII, and IFG composed a network, in which the functional integration of currently relevant sensory information into tactile processing circuits occured. Modeling the causal influences within this functional network using DCM revealed that IFG was intimately involved in

5.4. Discussion

Model	Modulatory input				Driving input	
	SI to SII	SII to IFG	SII to SI	IFG to SII	SI	IFG
1	A & C	A & C			S	
2	A & C	A & C	A & C	A & C	S	
3			A & C	A & C	S	
4	C	C	A	A	S	
5		C		A	S	
6	A & C	A & C			S	A
7	A & C	A & C	A & C	A & C	S	A
8			A & C	A & C	S	A
9	C	C	A	A	S	A
10		C		A	S	A

Table 5.3.: Model configurations. Models represent networks that consist of contralateral SI, SII, and IFG. Three different conditions act as driving input to areas and/or as modulatory influence on connectivity between areas: tactile stimulation (S), selective attention to either stimulus attribute (A), and change in the attended stimulus attribute (C). SI, primary somatosensory cortex; SII, secondary somatosensory cortex; IFG, inferior frontal gyrus.

the coordination of top-down attentional control and the processing of bottom-up sensory information.

In general, task performance involving selective attention to a specific tactile feature activated a widely distributed network including contralateral SI (areas 1, 2) and bilateral SII (OP 1, 4), aIPS (areas hIP2, hIP3), LPFC, IFG (area 44), as well as SMA (area 6), insular cortex, and the thalamus. This activation pattern altogether is in good agreement with the functional network active during tactile stimulation reported in Chapter 4. Passive tactile stimulation during the control task involved a similar network with smaller clusters of increased activity and additional responses in precuneus. The precuneus has been suggested to be an essential part of the default mode network active during the conscious resting state, in which people do not engage intentionally in sensory or motor activity (Cavanna 2007; Fransson and Marrelec 2008). The present study confirms this role of the precuneus in the default mode as it was activated during passive tactile stimulation, which did not require participants to process the somatosensory input actively.

The direct comparison between active attention and the control task revealed increased activity in SI (areas 1, 2) and SII (OP 1, 4), as well as in aIPS (areas hIP2, hIP3), IFG (area 44), insular cortex, SMA (area 6), and the thalamus during selective attention to

Chapter 5. Top-Down Attentional Bias for Gating Tactile Perception

any feature of the tactile stimulation independent of sensory input. In addition, analysis of the behavioral data revealed a significant reduction of the response times during both attention tasks compared with the control task, although tactile stimuli and behavioral responses were decoupled temporarily. These effects can be attributed to the engagement of selective attention, which, on the one hand, may result in reduced response times and increased accuracy in discrimination or detection tasks (e.g., Posner 1986; Sinclair et al. 2000), as decribed in Section 2.5. On the other hand, selective attention is known to enhance activity specifically in those areas that usually respond to the kind of sensory stimulation, which is attended to (Corbetta et al. 1991; Desimone and Duncan 1995; Burton et al. 1999). According to this, previous functional imaging work in humans revealed that tactile attention leads to increased activity in SI and SII (Mima et al. 1998; Burton et al. 1999; Johansen-Berg et al. 2000), which can be explained by resource mobilization due to the anticipation of incoming sensory input. Likewise, increased activity in inferior and superior parietal areas as well as in prefrontal cortex during selective attention to specific tactile features has been shown by previous functional imaging work (Van Boven et al. 2005; Burton et al. 2008). This task-dependent activation of the frontoparietal network during attentional task control has been described by many visual attention studies, and may be implicated in any goal-directed behavior involving cognitive selection of sensory information and responses (see Corbetta and Shulman 2002 for review).

The contrast between both attention tasks revealed increased BOLD responses in right IPC during P task performance, whereas we did not find any significant activations selective for the F task. Nonetheless, there were no significant differences between both tasks' accuracy levels and response times. Increased activity in IPC during pattern processing was already evident in the study of feature-specific representations presented in Chapter 4. Hegner and colleagues carried out a similar study, comparing tactile pattern and vibrotactile frequency discrimination (Hegner et al. 2010), in which activations in the right IPS, supramarginal gyrus in IPC, and SII, were significantly increased during the pattern task compared with the frequency task. No regions were found to be more activated during frequency than during pattern discrimination. The authors reasoned that different neuronal representations for the maintenance of pattern and frequency information might lead to the lack of regionally specific effects for the frequency task. This view is supported by a vibrotactile attention study, in which no regional differences were identified between frequency and duration discrimination of vibrotactile stimuli (Burton et al. 2008).

5.4. Discussion

Testing differential effects of change detection, we could show that correctly judged change in the attended stimulus attribute evoked robust activations in a frontoparietal network including bilateral IFG (area 44), aIPS (areas hIP2, hIP3), LPFC, and insular cortex, as well as right IPC. These effects were independent of change in the non-attended stimulus attribute as a comparison between trials with change in the non-attended stimulus attribute and trials without change in the non-attended feature revealed. Furthermore, employing a conjunction analysis of the contrasts (1) P task trials with change in pattern vs. F task trials with change in pattern and (2) F task trials with change in frequency vs. P task trials with change in frequency, we found increased activations in a similar network. This suggests that these effects were independent of sensory input, representing currently task-relevant information as the result of top-down directed attentional control. Supporting evidence comes from functional imaging work in the visual domain identifying specific responses to task-relevant information in a distributed network of frontal and parietal areas (Hon et al. 2006; Hampshire et al. 2007; Thompson and Duncan 2009). Using two related tasks, in which participants selectively attended either to one of two differentially colored shapes or to one of two different words presented simultaneously, Hon and colleagues showed strong frontoparietal activity associated with changes in the attended stimulus (Hon et al. 2006). These results point to selective representations of currently relevant information in regions of the frontal and parietal cortex, providing evidence for a link between conscious perception and frontoparietal activity. Interestingly, the activations reflected neither complex decision-making nor response selection, but simple update of attended visual information, as there was no task to perform and no behavior to control in this study. In line with this previous evidence, the present study revealed the representation of currently task-relevant information in a distributed network of frontal and parietal areas while performing a tactile attention task. These results corroborate that the mechansims for attentional task control in the somatosensory modality are similar to those identified in the visual domain.

Using PPI analysis we further studied task-induced changes in the effective connectivity of areas that were implicated in selective attention to tactile stimulus attributes. We identified a network composed of SI, SII, and IFG, in which the functional integration of currently relevant sensory information into tactile processing circuits appeared to occur. As part of the frontoparietal network the IFG in the human LPFC might play a prominent role in the cognitive control of action and attention. Consistent with this, category- or target-selective activity in this region has been observed in monkeys, using neurophysiological recordings (Freedman et al. 2001; Duncan 2001), and in humans, using functional imaging (Jiang et al. 2007; Hampshire et al. 2009). In addition, we found a representation of task-relevant frequency information in the prefrontal cortex during a delayed match-to-sample task in an accompanying study (Spitzer et al. 2010). This previous evidence indicates that a fundamental principle of prefrontal function might be the ability to adaptively represent specific information, which is relevant to current concerns. In particular, Hampshire and colleagues identified the IFG as selectively responsive to those stimuli that are most relevant to the currently intended task schema, independent of stimulus frequency, attentional load, and inhibitory processes (Hampshire et al. 2009). The latter component refers to the longstanding hypothesis that the IFG might play an important role in inhibitory control, resulting from studies employing variants of the go/no-go (GNG) or stop signal task (SST; Logan and Cowan 1984; Rubia et al. 2003; for review see Aron et al. 2004). This kind of tasks requires participants to suppress a routine response when an infrequent stop cue is presented. In a follow-up study addressing the question raised, the authors revealed that the IFG is not specifically involved in response inhibition but more generally in the detection of the task-relevant stop cue (Hampshire et al. 2010). This supports the view that the IFG can selectively be tuned to respond to sensory input, which is at the current focus of intention. On the other hand, there is further evidence for a crucial role of the IFG in the mediation of cognitive control, exerting effects by specifically modulating the processing in other brain areas (Staines et al. 2002; Li et al. 2007; Duann et al. 2009). Relating to this, Staines and colleagues identified the IFG in the human prefrontal cortex as likely candidate to mediate the modulation of early somatosensory areas in a vibrotactile attention task (Staines et al. 2002). Such modulation might occur by facilitating the sensitivity of neuronal networks in sensory areas for the accomplishment of the task, which required participants to sustain attention and to extract task-relevant features of the vibrotactile stimulation. The present evidence for the functional significance of the IFG both during selective attention to tactile features and during change detection complements the previous findings, rendering this area as likely candidate to act as the neural substrate for the coordination of top-down directed control and bottom-up sensory processing.

In addition to the present evidence for the role of a network composed of SI, SII, and IFG in task performance, our network analyses using DCM elucidated the causal influences of the experimental manipulations on effective connectivity within this network. Selective attention to either stimulus attribute significantly enhanced the top-down directed effec-

5.4. Discussion

tive connectivity between IFG and SII, whereas the change in the task-relevant stimulus attribute augmented the feedforward connection linking SII and IFG significantly. One potential interpretation of these results is that top-down directed signals originating in IFG might sensitize somatosensory areas for the incoming input, which requires these areas to correctly extract the task-relevent stimulus attribute and to detect the respective feature change. In turn, information about the detection of task-relevant changes is then transmitted to the IFG in order to induce further higher-order processing such as subjective awareness and response selection. This interpretation is in line with previous reports of top-down mechanisms arising in prefrontal cortex that may drive responses in sensory areas responsible for collecting evidence about the presence of task-relevant stimuli in a visual detection task (Summerfield et al. 2006), or that are even able to generate mental images in category-specific areas in a visual imagery task (Mechelli et al. 2004). The significance of the IFG in this function in the present experimental context may emphasize that this area's role generalizes across the coordination of bottom-up sensory processing and top-down directed attentional control indicated by previous evidence (e.g., Staines et al. 2002; Hampshire et al. 2009; Duann et al. 2009; Asplund et al. 2010). Given these conclusions, the present findings contribute to the delineation of frontoparietal function, in particular with respect to functional interactions between components of the frontoparietal network and sensory areas that create the prerequisites for task accomplishment.

In sum, the findings presented here revealed distinct neuronal mechanisms that mediate context-related activity in the human brain during a tactile detection task, which required the deployment of top-down directed selective attention and the extraction of task-relevant sensory information. As part of the frontoparietal network, which is implicated in cognitive control of action and attention, the IFG in the human prefrontal cortex may exert top-down control to modulate the processing in somatosensory areas responsible for identifying the task-relevant feature of the tactile stimulation. In addition, this area may contribute to the bottom-up transmission of the extracted task-relevant information into a widely distributed frontoparietal network selectively activated during the detection of those aspects of the sensory stimulation that are at the current focus of attention. It thereby seems that the neuronal mechanisms identified here may generalize across modalities and cognitive demands, indicating a central role of the frontoparietal network, and in particular the IFG, in the cognitive coordination of top-down attentional control and bottom-up sensory processing.

Chapter 5. Top-Down Attentional Bias for Gating Tactile Perception

Chapter 6.
Summary and Model Extension

The aim of the present thesis was to extend the model of somatosensory processing for perception and action proposed by Dijkerman and de Haan (Dijkerman and de Haan 2007), focussing on tactile feature processing and attentional modulation during tactile perception. Both aspects have attached little importance in the original model. Two experimental investigations were carried out, in which tactile stimulation was applied to the index finger tip of the participants' hand, and specific stimulus attributes were associated with different cognitive demands. Using fMRI we sought to determine the role of feature-specific higher-order processing for tactile perception but also the functional significance of attentional top-down modulation during tactile task accomplishment.

In this chapter, following a brief summary of the findings presented in this thesis, we reconsider the model of somatosensory processing for perception and action introduced in Section 2.3. We then propose an extended version incorporating the results of the two experiments and give an outlook on future investigations that may build on the model.

6.1. Summary of Experimental Results

In the first experimental investigation, we used fMRI to examine the neuronal correlates associated with feature-specific processing of tactile stimulus attributes in humans under tightly controlled stimulation conditions. Different types of dynamic stimuli created the sensation of moving or stationary bar patterns during passive touch. Activity in somatosensory cortex was increased during both motion and pattern processing and modulated by motion directionality in SI and SII as well as by pattern orientation in aIPS. Furthermore, tactile motion and pattern processing induced activity in hMT+/V5, an area traditionally associated with visual motion perception, and in IPC, involving parts

Chapter 6. Summary and Model Extension

of the supramarginal and angular gyri. These responses covaried with subjects' individual perceptual performance in identifying the respective stimulus attribute, suggesting that hMT+/V5 and IPC contribute to conscious perception of specific tactile stimulus features. In addition, an analysis of effective connectivity using PPI models revealed increased functional coupling between SI and hMT+/V5 during motion processing, as well as between SI and IPC during pattern processing. This connectivity pattern provides evidence for the direct engagement of feature-specific cortical areas in tactile processing during somesthesis.

In the second study presented in this thesis we took advantage of fMRI to investigate the cortical network engaged in a tactile detection task that involved selective attention to a specific tactile stimulus attribute. The participants were asked to attend either to the spatial pattern or to the temporal frequency of complex Braille-like tactile stimuli and to detect changes in the respective stimulus attribute. Assessing task-dependent activations revealed a distributed network of somatosensory as well as prefrontal and parietal areas active during tactile task performance. Moreover, frontoparietal components of this task-related network including IFG, LPFC, and aIPS selectively responded to the detection of change in the task-relevant stimulus attribute. An analysis of effective connectivity using PPI modeling further revealed that SI, SII, and IFG composed a network related to the functional integration of task-relevant sensory information into tactile processing circuits. Modeling context-dependent causal influences within this functional network using DCM identified the crucial role of IFG in the cognitive coordination of top-down attentional control and the processing of bottom-up sensory information.

6.2. The Extended Model of Somatosensory Processing

In Section 2.3 of the present thesis, we introduced the model of somatosensory processing for perception and action that was initially proposed by Dijkerman and de Haan (Dijkerman and de Haan 2007). According to this model, a cortical pathway, which is responsible for conscious somatosensory perception and recognition of tactile objects, projects from SI via SII to the insula. Thereby, the PPC contributes to the spatio-temporal integration of the input. Information about an object's properties is provided by higher-order association areas. The authors regarded this processing stream as the somatosensory equivalent of the ventral "what" pathway in the visual system. A second route running from SI, either directly or via SII, to the PPC was suggested to subserve

6.2. The Extended Model of Somatosensory Processing

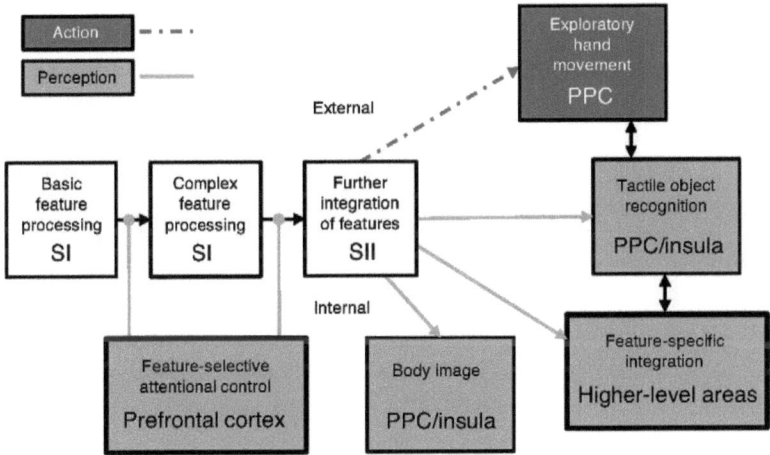

Figure 6.1.: Extended model of somatosensory processing for perception and action. The extended version of the model shown in Figure 2.2 incorporates the results of the two studies forming the basis for the present thesis. The prefrontal cortex may exert top-down directed attentional control over processing in somatosensory areas to guide tactile perception (left highlighted box; see Chapter 5), which may involve feature-specific integration in higher-level areas (right highlighted box; see Chapter 4). SI, primary somatosensory cortex; SII, secondary somatosensory cortex; PPC, posterior parietal cortex.

somatosensory processing for the guidance of action. This cortical pathway might constitute the tactile dorsal stream involved in somatosensory processing of "where" and "how". A further distinction, which is not possible in the visual system, was made between somatosensory processing of external stimuli and internal targets. In addition, tactile perceptual recognition operates in close cooperation with action-related mechanisms (e.g., exploratory hand movements) and with bodily awareness involving proprioceptive information (i.e., body image) in the somatosensory system. Therefore, the distinctions made between somatosensory processes for perception and action as well as between processing of external and internal targets are less independent than the "what" vs. "where" distinction proposed for the visual system. This is in particular demonstrated by the "dual" role of the PPC, contributing both to somatosensory processing of "what" for tactile perception and "how" for the guidance of action.

Chapter 6. Summary and Model Extension

Based on the results of the experimental investigations presented in Chapters 4 and 5 of this thesis, we extended the model of somatosensory processing for perception and action. Our focus was on somatosensory processing for conscious perception of external tactile input and, in particular, the role of feature-specific processing and attentional top-down modulation for tactile perception. Figure 6.1 illustrates the extended model, which now includes functions for feature-specific integration and feature-selective attentional control. The specific contributions are described in the following.

The Role of Feature-Specific Processing for Tactile Perception

In the first study presented in this thesis, we have shown that conscious perception of tactile stimulus attributes may engage feature-specific integration in higher-level cortical areas, which may be independent of the input's sensory modality. Based on these findings, a function for feature-specific integration in higher-level areas was added to the model of somatosensory processing for perception and action. This extension is illustrated by the right highlighted box in Figure 6.1 and by the pathways connecting somatosensory cortex, higher-level areas, and the PPC. Modality-specific somatosensory areas may interact with regions that are specifically involved in the integration of perceptual features such as motion or pattern into a conscious perceptual concept. Such higher-order processing of individual stimulus features in modality-independent areas might be arranged in parallel to the processing pathways from somatosensory cortex to the insula and the PPC. The integration of feature-specific information into a coherent representation of the tactile input might further require temporal and spatial processing, which is subserved by the PPC. Such functional interplay between higher-level areas and the PPC for tactile perception is illustrated by the bidirectional link connecting these processes in Figure 6.1. In line with this proposal, we provided evidence for motion-specific responses in area hMT+/V5 in Chapter 4 and showed that the functional integration of this motion representation occured in a network of SI, hMT+/V5, and aIPS in the PPC during tactile motion processing. We further identified the IPC as selectively responsive to pattern information. We did not directly show the involvement of SII in these networks but provided suggestive evidence for this assumption. Therefore, the proposed processing stream for feature-specific integration might run from SI either directly or via SII to higher-level cortical areas. As emphasized several times, this functional segregation does not necessarily preclude close interactions between feature-specific areas, insula, and PPC, as tactile perception and object recognition require the integration of cutaneous, proprioceptive, and spatio-temporal information. Another candidate for such feature-specific integration of multisensory in-

6.2. The Extended Model of Somatosensory Processing

put might be the LOC, as discussed in Section 4.4. Several studies have shown that the LOC is activated during the processing of visual and tactile information about an object's shape (e.g., Amedi et al. 2001; Pietrini et al. 2004; Lucan et al. 2010). The LOC therefore appears to constitute a modality-independent area related to higher-order perceptual representations of objects. Relating to this, Lacey and Sathian proposed a model of multisensory object recognition, in which the LOC contains a representation of object form that can be flexibly accessed either bottom-up or top-down, independently of the input modality (Lacey and Sathian 2011). Unisensory representations feed forward into this multisensory representation that supports higher-order object recognition. In this way, the processing of sensory input from any specific modality involves general higher-order representations of task-relevant stimulus features. Our findings support such an integrative view, indicating that modality-specific representations of tactile information may be closely related to feature-specific integration in modality-independent areas. The present and the previous evidence for the recruitment of higher-order cortical areas involved in multisensory integration suggest that feature-specific higher-order processing might play a crucial role for tactile perception and object recognition.

The Role of Attentional Modulation for Tactile Perception

The second experimental investigation revealed distinct neuronal mechanisms that mediate context-related activity in the human brain during a tactile change-detection task. The human prefontal cortex may exert attentional control over processing in somatosensory areas to guide selective perception of task-relevant stimulus attributes. Based on our results, we added a function for attentional top-down modulation to the model of somatosensory processing for perception and action. An additional box in the model schema represents this extension (left highlighted box in Figure 6.1), including the signals from prefrontal cortex that modulate the processing streams between somatosensory areas. We subdivided the box devoted to SI in the former version of the model in order to illustrate that each of the different steps of feature processing may be modulated by feature-selective attention. Somatosensory areas are sensitized about the anticipated input by means of top-down directed attentional modulation. This mechanism allows for the identification, the extraction, and further processing of task-relevant stimulus features. Successfully extracted information, which is at the current focus of intention, can then be transmitted to feature-specific and/or posterior parietal areas. These areas further integrate this information into a conscious perceptual representation, which is required for task accomplishment. In Chapter 5, we could show that the IFG in the human prefrontal

Chapter 6. Summary and Model Extension

cortex may play a crucial role in mediating top-down attentional control over somatosensory areas during tactile task performance. As part of the frontoparietal network involved in cognitive control of action and attention, the IFG may send top-down signals to modulate the processing in somatosensory areas that are in charge of task-relevant feature selection. The feedback connection from prefrontal cortex to somatosensory areas may not necessarily be unidirectional, as the prefrontal cortex has to be updated with sensory information in the service of attentional control. In addition, the IFG may further participate in the bottom-up transmission of the extracted information to the PPC. In this context, we provided evidence that the connection from SII to IFG was modulated by the presence of change in the task-relevant stimulus attribute. Furthermore, the aIPS in the PPC as part of the frontoparietal network was selectively activated by the detection of change in the task-relevant stimulus attribute. Thus, our findings may suggest a central role of the frontoparietal network, and in particular the IFG, in the functional interplay of top-down attentional control and bottom-up sensory processing, which may generalize across modalities and cognitive demands. The concept of attentional top-down modulation further relates to the theory of predictive coding (Rao and Ballard 1999) and the importance of expectation for sensory processing. Expectation reflects prior information such as knowledge or experience about what is possible or plausible in the forthcoming perception (reviewed in Summerfield and Egner 2009). According to this, prior information acts as top-down directed influence on sensory areas during perceptual inference, which may guide perception and in doing so facilitates the interpretation of sensory input. However, the understanding of how expectation and attention are related, how they may differ and how they interact in terms of top-down modulation remains to be deepened further in future investigations.

6.3. Outlook

Overall, the postulated model of somatosensory processing for perception and action incorporates present and previous evidence delineating the cortical processing of somatosensory information during conscious perceptual inference. The model was revisited in the light of new insights and extended based on the findings presented in this thesis. It constitutes a description of how the human somatosensory system is organized to subserve conscious perception of external tactile input. Of course, tactile performance involving perceptual inference of external objects requires close interactions between action-related and

6.3. Outlook

perception-related processes, as well as the coordinated processing of internal and external information. The model was intended to enable the formulation of testable hypotheses in order to advance the understanding of how the somatosensory system manages conscious perception. In this way, the extended model complements the current knowledge of somatosensory processing for conscious perception and contributes to the understanding of the human somatosensory system. In addition to its scientific value, the knowledge derived in this way may implicate clinical applications in terms of neurorehabilitation after peripheral nerve injury or stroke. Basic science research is important as foundation for the development of innovative technologies that substitute for lost abilities in humans following injury or disease. Applications such as neuroprosthetic devices controlled by the brain are becoming a realistic possibility for restoring lost sensory or motor function. The elucidation of processes and neural substrates involved in somatosensory processing is now making progress but it will require multidisciplinary investigations aiming at delineating interactions between the functional entities proposed in the model. Only such cooperation will lead to a better understanding of the somatosensory system, according to Seneca's social rule that "one hand washes the other".

Chapter 6. Summary and Model Extension

Chapter 7.
Conclusions

In the present thesis, we proposed an extended version of the model of somatosensory processing for perception and action introduced by Dijkerman and de Haan (Dijkerman and de Haan 2007). The suggested extension predominantly concerned the cortical pathway, which is responsible for somatosensory processing for tactile perception and object recognition, as proposed by the authors. In two experimental investigations using fMRI, we sought to determine the role of feature-specific processing and attentional top-down modulation for tactile perception. The findings presented in this thesis were incorporated as organizing principles of perceptual inference into the model of somatosensory processing for perception and action, and led to the following conclusions:

- Feature-specific integration in higher-order areas aids conscious perception of tactile stimulus attributes.

- Top-down modulation of somatosensory areas helps focussing attentional resources on task-relevant aspects of the tactile input.

Both feature-specific and top-down processing gained in importance over the last several years. Regarding the former, there is increasing evidence for the close collaboration between the sensory modalities, especially between the somatosensory and the visual system. As an example, the act of manipulating objects usually requires visual guidance towards the object but also somatosensory feedback to inform us about the object's textural and spatial properties. Close interactions between the modalities allow us to fine-tune the forces applied to the object. As illustrated by this example, the functional interplay between modalities as well as feature-specific modality-independent processing expand the individual capabilities of each system component. Consistent with this, we provided

Chapter 7. Conclusions

evidence supporting the notion that feature-specific processing might be a general rule during perceptual inference.

Concerning top-down processing, there is strong evidence that top-down attentional control may shape the way, in which sensory inputs are being processed and perceived. This relates to the hypothesis that sensory systems are limited in their capacity of processing incoming information. Top-down mechanisms provide the organism with the ability to selectively suppress irrelevant information and enhance the representations of those aspects of the input that are of current relevance. Several studies have directed considerable effort towards elucidating these mechanisms of selective attention in the visual system. In the present thesis, we demonstrated that the somatosensory system likewise engages top-down processing for the guidance of selective perception of task-relevant stimulus attributes, suggesting that these mechanisms may generalize across modalities and cognitive demands.

Appendix A.
BOLD Time Courses

Statistical maps of functional neuroimaging data give a fair impression of the regional significance and the spatial extent of an experimental effect but they are also just thresholded images, which in many cases represent relative differences in signal intensity and often do not convey the real nature of the effect. In order to allow for proper evaluation of evoked activation profiles, it is therefore helpful to extract and to investigate the original imaging data, referred to as BOLD time courses.

BOLD time courses were generated using the rfxplot toolbox (Gläscher 2009) for SPM8 (Wellcome Department of Imaging Neuroscience, University College London, UK). For every subject, the data were extracted from voxels showing the individual activation maximum within a sphere of 5 mm radius constructed around the group-level activation maximum. Averaged across the group, the mean-corrected BOLD time courses were plotted time-locked to tactile stimulation onset (stimulus duration was 4 s). The errors plotted as dotted lines around the mean response are ± 1 standard error of the mean.

Appendix A. BOLD Time Courses

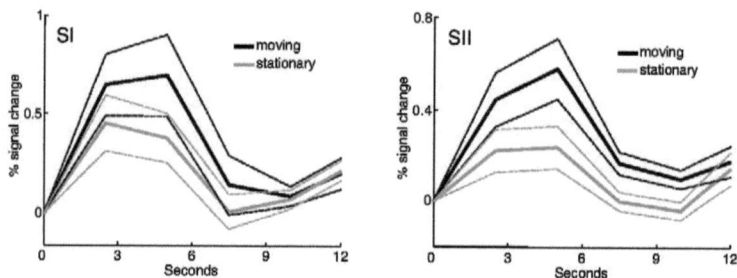

Figure A.1.: Group-averaged BOLD time courses extracted from SI (left) and SII (right) for moving and stationary trials.

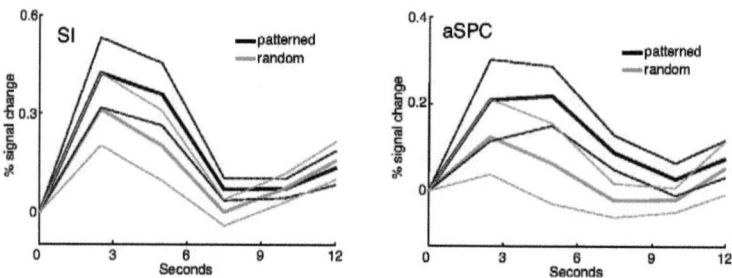

Figure A.2.: Group-averaged BOLD time courses extracted from SI (left) and aSPC (right) for patterned and random trials.

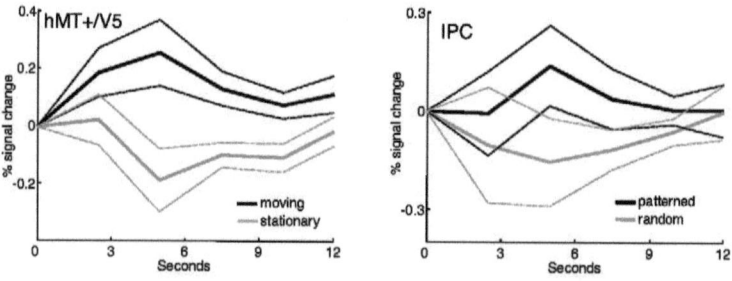

Figure A.3.: Group-averaged BOLD time courses extracted from hMT+/V5 (left) and IPC (right) for moving/stationary trials and for patterned/random trials.

Appendix B.

Additional Figures

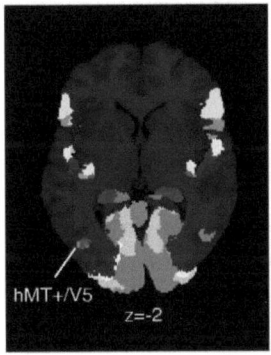

Figure B.1.: Overlap with anatomically defined map for hMT+/V5 activation. The part of the activation for moving vs. stationary stimuli that overlaps with the anatomically defined map for hMT+/V5 is superimposed on the probabilistic atlas provided by the Anatomy toolbox for SPM.

Appendix B. Additional Figures

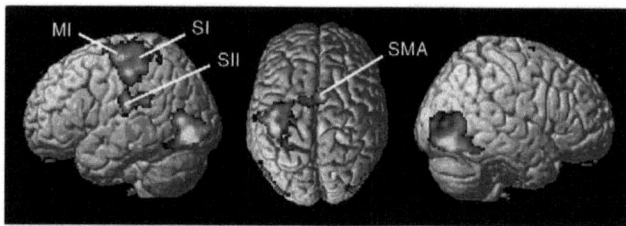

Figure B.2.: Functional regions active during motor responses. Contrasting response onsets with baseline revealed somatomotor-related areas in the left hemisphere (contralateral to the hand holding the response keys) including MI in the precentral gyrus as well as SI, SII, and SMA. Lateral occipital activations can be attributed to the visual cue advising participants to press the response button. Statistical analysis involved a one-sided t-test at the group level using a significance threshold of $p_{cluster} < 0.05$, corrected for multiple comparisons. SI, primary somatosensory cortex; SII, secondary somatosensory cortex; MI, primary motor cortex; SMA, supplementary motor area.

Appendix C.
Bootstrapping Correlations

In order to assess whether the positive relationships determined between the BOLD responses in hMT+/V5 (IPC) and participants' accuracy in identifying moving (patterned) stimuli correctly (see Section 4.3) were indeed reliable, bootstrap sampling was performed (Efron and Tibshirani 1986). The bootstrap method involves choosing random samples with replacement from a given data set and analyzing each sample the same way. Sampling with replacement means that every sample is returned to the data set for the next sampling, so that a specific data point from the original data set might appear multiple times in a bootstrapped sample. The range of sample estimates obtained allows establishing the uncertainty of the quantity that is estimated.

Using MATLAB (v7.5, The MathWorks, Natick, MA) one thousand bootstrapped samples each were drawn with replacement from the two original data sets. Samples were drawn pairwise, meaning that when a BOLD response in hMT+/V5 (IPC) was sampled, the corresponding performance value of identifying moving (patterned) stimuli correctly was also drawn. For each bootstrapped sample the correlation coefficient was calculated. The histograms in Figure C.1 show the variation of the correlation coefficients across all bootstrap samples for (a) the relationship between BOLD responses in hMT+/V5 and participants' performance in identifying moving stimuli correctly and (b) the relationship between BOLD responses in IPC and participants' performance in identifying patterned stimuli correctly. The mean values of the correlation coefficients were 0.70 for the performance-dependent covariation in hMT+/V5 and 0.63 for the performance-dependent covariation in IPC.

To test whether the correlations were significantly different from zero, one-sided confidence intervals were determined. After sorting both bootstrap samples, the fiftieth largest values were 0.3894 for the covariation in hMT+/V5 and 0.2026 for the covariation in IPC.

Appendix C. Bootstrapping Correlations

Both values exceed zero, which indicates that both correlations were significantly different from zero at a threshold of $p<0.05$. The tenth largest values were 0.1717 and -0.2698 for the covariation in hMT+/V5 and IPC, respectively. Thus, the correlation in hMT+/V5 was even significant at $p<0.01$ suggesting that the performance-dependent covariation in hMT+/V5 was more reliable than in IPC.

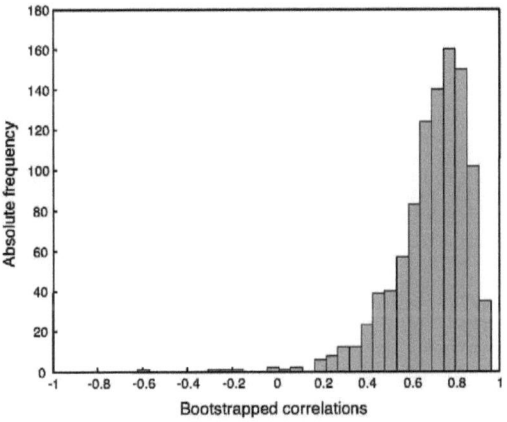

(a) Performance-dependent covariation in hMT+/V5

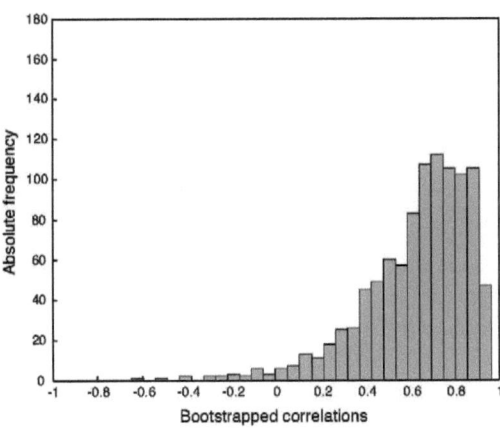

(b) Performance-dependent covariation in IPC

Figure C.1.: Bootstrapping correlations between (a) BOLD responses in area hMT+/V5 and participants' accuracy in identifying moving stimuli correctly and (b) BOLD responses in IPC and participants' accuracy in identifying patterned stimuli correctly.

Appendix C. Bootstrapping Correlations

Bibliography

Amedi A, Malach R, Hendler T, Peled S, and Zohary E (2001). Visuo-haptic object-related activation in the ventral visual pathway. *Nat Neurosci*, 4:324–330.

Aron AR, Robbins TW, and Poldrack RA (2004). Inhibition and the right inferior frontal cortex. *Trends Cogn Sci*, 8:170–177.

Ashburner J, Chen CC, Flandin G, Henson RN, Kiebel SJ, Kilner JM, Litvak V, Moran R, Penny WD, Stephan KE, Hutton C, Glauche V, Mattout J, and Phillips C (2009). *SPM8 Manual*. Functional Imaging Laboratory, University College London.

Asplund CL, Todd JJ, Snyder AP, and Marois R (2010). A central role for the lateral prefrontal cortex in goal-directed and stimulus-driven attention. *Nat Neurosci*, 13:507–512.

Beauchamp MS, Yasar NE, Kishan N, and Ro T (2007). Human MST but not MT responds to tactile stimulation. *J Neurosci*, 27:8261–8267.

Bell J, Bolanowski S, and Holmes MH (1994). The structure and function of Pacinian corpuscles: a review. *Prog Neurobiol*, 42:79–128.

Bensmaia SJ, Denchev PV, Dammann JF 3rd, Craig JC, and Hsiao SS (2008). The representation of stimulus orientation in the early stages of somatosensory processing. *J Neurosci*, 28:776–786.

Binkofski F, Buccino G, Posse S, Seitz RJ, Rizzolatti G, and Freund H (1999). A fronto-parietal circuit for object manipulation in man: evidence from an fMRI-study. *Eur J Neurosci*, 11:3276–3286.

Blake R, Sobel KV, and James TW (2004). Neural synergy between kinetic vision and touch. *Psychol Sci*, 15:397–402.

Blankenburg F, Ruff CC, Deichmann R, Rees G, and Driver J (2006). The cutaneous rabbit illusion affects human primary sensory cortex somatotopically. *PLoS Biol*, 4:e69.

Bodegård A, Geyer S, Grefkes C, Zilles K, and Roland PE (2001). Hierarchical processing of tactile shape in the human brain. *Neuron*, 31:317–328.

Bodegård A, Geyer S, Naito E, Zilles K, and Roland PE (2000). Somatosensory areas in man activated by moving stimuli: cytoarchitectonic mapping and PET. *Neuroreport*, 11:187–191.

Bremmer F, Schlack A, Shah NJ, Zafiris O, Kubischik M, Hoffmann K, Zilles K, and Fink GR (2001). Polymodal motion processing in posterior parietal and premotor cortex: a human fMRI study strongly implies equivalencies between humans and monkeys. *Neuron*, 29:287–296.

Brodmann K (1909). *Vergleichende Lokalisationslehre der Großhirnrinde: in ihren Principien dargestellt auf Grund des Zellenbaues*. Leipzig: Johann Ambrosius Barth Verlag. 335 p.

Burton H, Abend NS, MacLeod AM, Sinclair RJ, Snyder AZ, and Raichle ME (1999). Tactile attention tasks enhance activation in somatosensory regions of parietal cortex: a positron emission tomography study. *Cereb Cortex*, 9:662–674.

Burton H, Sinclair RJ, and McLaren DG (2008). Cortical network for vibrotactile attention: a fMRI study. *Hum Brain Mapp*, 29:207–221.

Cavanna AE (2007). The precuneus and consciousness. *CNS Spectr*, 12:545–552.

Corbetta M, Miezin FM, Dobmeyer S, Shulman GL, and Petersen SE (1991). Selective and divided attention during visual discriminations of shape, color, and speed: functional anatomy by positron emission tomography. *J Neurosci*, 11:2383–2402.

Corbetta M and Shulman GL (2002). Control of goal-directed and stimulus-driven attention in the brain. *Nat Rev Neurosci*, 3:201–215.

Costanzo RM and Gardner EP (1980). A quantitative analysis of responses of direction-sensitive neurons in somatosensory cortex of awake monkeys. *J Neurophysiol*, 43:1319–1341.

Bibliography

Culham JC, Cavina-Pratesi C, and Singhal A (2006). The role of parietal cortex in visuomotor control: what have we learned from neuroimaging? *Neuropsychologia*, 44:2668–2684.

Culham JC and Kanwisher NG (2001). Neuroimaging of cognitive functions in human parietal cortex. *Curr Opin Neurobiol*, 11:157–163.

Cusick CG, Wall JT, Felleman DJ, and Kaas JH (1989). Somatotopic organization of the lateral sulcus of owl monkeys: area 3b, S-II, and a ventral somatosensory area. *J Comp Neurol*, 282:169–190.

Darian-Smith I, Goodwin A, Sugitani M, and Heywood J (1984). The tangible features of textured surfaces: their representation in the monkey's somatosensory cortex. In *Dynamic Aspects of Neocortical Function*. New York: John Wiley & Sons Ltd.

Deibert E, Kraut M, Kremen S, and Hart J Jr (1999). Neural pathways in tactile object recognition. *Neurology*, 52:1413–1417.

Deshpande G, Hu X, Lacey S, Stilla R, and Sathian K (2010). Object familiarity modulates effective connectivity during haptic shape perception. *Neuroimage*, 49:1991–2000.

Deshpande G, Hu X, Stilla R, and Sathian K (2008). Effective connectivity during haptic perception: a study using Granger causality analysis of functional magnetic resonance imaging data. *Neuroimage*, 40:1807–1814.

Desimone R and Duncan J (1995). Neural mechanisms of selective visual attention. *Annu Rev Neurosci*, 18:193–222.

DiCarlo JJ and Johnson KO (2000). Spatial and temporal structure of receptive fields in primate somatosensory area 3b: effects of stimulus scanning direction and orientation. *J Neurosci*, 20:495–510.

DiCarlo JJ and Johnson KO (2002). Receptive field structure in cortical area 3b of the alert monkey. *Behav Brain Res*, 135:167–178.

Dijkerman HC and de Haan EH (2007). Somatosensory processes subserving perception and action. *Behav Brain Sci*, 30:189–201.

Bibliography

Disbrow E, Roberts T, and Krubitzer L (2000). Somatotopic organization of cortical fields in the lateral sulcus of *Homo sapiens*: evidence for SII and PV. *J Comp Neurol*, 418:1–21.

Dodson MJ, Goodwin AW, Browning AS, and Gehring HM (1998). Peripheral neural mechanisms determining the orientation of cylinders grasped by the digits. *J Neurosci*, 18:521–530.

Duann JR, Ide JS, Luo X, and Li CS (2009). Functional connectivity delineates distinct roles of the inferior frontal cortex and presupplementary motor area in stop signal inhibition. *J Neurosci*, 29:10171–10179.

Dubner R and Zeki SM (1971). Response properties and receptive fields of cells in an anatomically defined region of the superior temporal sulcus in the monkey. *Brain Res*, 35:528–532.

Duncan J (2001). An adaptive coding model of neural function in prefrontal cortex. *Nat Rev Neurosci*, 2:820–829.

Duncan J (2010). The multiple-demand (MD) system of the primate brain: mental programs for intelligent behaviour. *Trends Cogn Sci*, 14:172–179.

Duncan J and Owen AM (2000). Common regions of the human frontal lobe recruited by diverse cognitive demands. *Trends Neurosci*, 23:475–483.

Efron B and Tibshirani R (1986). Bootstrap methods for standard errors, confidence intervals, and other measures of statistical accuracy. *Statist Sci*, 1:54–77.

Eickhoff SB, Amunts K, Mohlberg H, and Zilles K (2006a). The human parietal operculum. II. Stereotaxic maps and correlation with functional imaging results. *Cereb Cortex*, 16:268–279.

Eickhoff SB, Jbabdi S, Caspers S, Laird AR, Fox PT, Zilles K, and Behrens TE (2010). Anatomical and functional connectivity of cytoarchitectonic areas within the human parietal operculum. *J Neurosci*, 30:6409–6421.

Eickhoff SB, Schleicher A, Zilles K, and Amunts K (2006b). The human parietal operculum. I. Cytoarchitectonic mapping of subdivisions. *Cereb Cortex*, 16:254–267.

Eickhoff SB, Stephan KE, Mohlberg H, Grefkes C, Fink GR, Amunts K, and Zilles K (2005). A new SPM toolbox for combining probabilistic cytoarchitectonic maps and functional imaging data. *Neuroimage*, 25:1325–1335.

Fitzgerald PJ, Lane JW, Thakur PH, and Hsiao SS (2006). Receptive field properties of the macaque second somatosensory cortex: representation of orientation on different finger pads. *J Neurosci*, 26:6473–6484.

Fransson P and Marrelec G (2008). The precuneus/posterior cingulate cortex plays a pivotal role in the default mode network: evidence from a partial correlation network analysis. *Neuroimage*, 42:1178–1184.

Freedman DJ, Riesenhuber M, Poggio T, and Miller EK (2001). Categorical representation of visual stimuli in the primate prefrontal cortex. *Science*, 291:312–316.

Freeman J, Brouwer GJ, Heeger DJ, and Merriam EP (2011). Orientation decoding depends on maps, not columns. *J Neurosci*, 31:4792–4804.

Friedman DP, Murray EA, O'Neill JB, and Mishkin M (1986). Cortical connections of the somatosensory fields of the lateral sulcus of macaques: evidence for a corticolimbic pathway for touch. *J Comp Neurol*, 252:323–347.

Friston KJ (1995). Functional and effective connectivity in neuroimaging: a synthesis. *Hum Brain Mapp*, 2:56–78.

Friston KJ, Ashburner JT, Kiebel S, Nichols TE, and Penny WD, editors (2006). *Statistical Parametric Mapping: The Analysis of Functional Brain Images*. London: Academic Press. 656 p.

Friston KJ, Büchel C, Fink GR, Morris J, Rolls E, and Dolan RJ (1997). Psychophysiological and modulatory interactions in neuroimaging. *Neuroimage*, 6:218–229.

Friston KJ, Frith CD, Liddle PF, and Frackowiak RS (1991). Comparing functional (PET) images: the assessment of significant change. *J Cereb Blood Flow Metab*, 11:90–99.

Friston KJ, Harrison L, and Penny WD (2003). Dynamic causal modelling. *Neuroimage*, 19:1273–1302.

Friston KJ, Holmes AP, Worsley KJ, Poline J, Frith C, and Frackowiak RS (1995). Statistical parametric maps in functional imaging: a general linear approach. *Hum Brain Mapp*, 2:189–210.

Friston KJ, Mechelli A, Turner R, and Price CJ (2000). Nonlinear responses in fMRI: the Balloon model, Volterra kernels, and other hemodynamics. *Neuroimage*, 12:466–477.

Gläscher J (2009). Visualization of group inference data in functional neuroimaging. *Neuroinformatics*, 7:73–82.

Goebel R, Khorram-Sefat D, Muckli L, Hacker H, and Singer W (1998). The constructive nature of vision: direct evidence from functional magnetic resonance imaging studies of apparent motion and motion imagery. *Eur J Neurosci*, 10:1563–1573.

Grefkes C and Fink GR (2005). The functional organization of the intraparietal sulcus in humans and monkeys. *J Anat*, 207:3–17.

Grefkes C, Weiss PH, Zilles K, and Fink GR (2002). Crossmodal processing of object features in human anterior intraparietal cortex: an fMRI study implies equivalencies between humans and monkeys. *Neuron*, 35:173–184.

Hadjikhani N and Roland PE (1998). Cross-modal transfer of information between the tactile and the visual representations in the human brain: a positron emission tomographic study. *J Neurosci*, 18:1072–1084.

Hagen MC, Franzén O, McGlone F, Essick G, Dancer C, and Pardo JV (2002). Tactile motion activates the human middle temporal/V5 (MT/V5) complex. *Eur J Neurosci*, 16:957–964.

Hampshire A, Chamberlain SR, Monti MM, Duncan J, and Owen AM (2010). The role of the right inferior frontal gyrus: inhibition and attentional control. *Neuroimage*, 50:1313–1319.

Hampshire A, Duncan J, and Owen AM (2007). Selective tuning of the blood oxygenation level-dependent response during simple target detection dissociates human frontoparietal subregions. *J Neurosci*, 27:6219–6223.

Hampshire A, Thompson R, Duncan J, and Owen AM (2009). Selective tuning of the right inferior frontal gyrus during target detection. *Cogn Affect Behav Neurosci*, 9:103–112.

Haynes JD and Rees G (2005). Predicting the orientation of invisible stimuli from activity in human primary visual cortex. *Nat Neurosci*, 8:686–691.

Hegner YL, Lee Y, Grodd W, and Braun C (2010). Comparing tactile pattern and vibrotactile frequency discrimination: a human fMRI study. *J Neurophysiol*, 103:3115–3122.

Hinkley LB, Krubitzer LA, Padberg J, and Disbrow EA (2009). Visual-manual exploration and posterior parietal cortex in humans. *J Neurophysiol*, 102:3433–3446.

Hollins M, Bensmaia SJ, and Roy EA (2002). Vibrotaction and texture perception. *Behav Brain Res*, 135:51–56.

Hon N, Epstein RA, Owen AM, and Duncan J (2006). Frontoparietal activity with minimal decision and control. *J Neurosci*, 26:9805–9809.

Hsiao SS, Lane J, and Fitzgerald P (2002). Representation of orientation in the somatosensory system. *Behav Brain Res*, 135:93–103.

Hsiao SS, O'Shaughnessy DM, and Johnson KO (1993). Effects of selective attention on spatial form processing in monkey primary and secondary somatosensory cortex. *J Neurophysiol*, 70:444–447.

Huettel SA, Song AW, and McCarthy G (2009). *Functional Magnetic Resonance Imaging*. Sunderland, MA: Sinauer Associates. 510 p.

Jiang X, Bradley E, Rini RA, Zeffiro T, Vanmeter J, and Riesenhuber M (2007). Categorization training results in shape- and category-selective human neural plasticity. *Neuron*, 53:891–903.

Johansen-Berg H, Christensen V, Woolrich M, and Matthews PM (2000). Attention to touch modulates activity in both primary and secondary somatosensory areas. *Neuroreport*, 11:1237–1241.

Johnson KO (2001). The roles and functions of cutaneous mechanoreceptors. *Curr Opin Neurobiol*, 11:455–461.

Johnson KO and Hsiao SS (1992). Neural mechanisms of tactual form and texture perception. *Annu Rev Neurosci*, 15:227–250.

Bibliography

Johnson KO, Yoshioka T, and Vega-Bermudez F (2000). Tactile functions of mechanoreceptive afferents innervating the hand. *J Clin Neurophysiol*, 17:539–558.

Kaas JH (1983). What, if anything, is SI? Organization of first somatosensory area of cortex. *Physiol Rev*, 63:206–231.

Kaas JH, Nelson RJ, Sur M, Lin CS, and Merzenich MM (1979). Multiple representations of the body within the primary somatosensory cortex of primates. *Science*, 204:521–523.

Kamitani Y and Tong F (2005). Decoding the visual and subjective contents of the human brain. *Nat Neurosci*, 8:679–685.

Kamitani Y and Tong F (2006). Decoding seen and attended motion directions from activity in the human visual cortex. *Curr Biol*, 16:1096–1102.

Kandel ER, Schwartz JH, and Jessell TM (2000). *Principles of Neural Science*. New York: McGraw-Hill. 1568 p.

Kass RE and Raftery AE (1995). Bayes factors. *J Am Stat Assoc*, 90:773–795.

Khalsa PS, Friedman RM, Srinivasan MA, and Lamotte RH (1998). Encoding of shape and orientation of objects indented into the monkey fingerpad by populations of slowly and rapidly adapting mechanoreceptors. *J Neurophysiol*, 79:3238–3251.

Kitada R, Kito T, Saito DN, Kochiyama T, Matsumura M, Sadato N, and Lederman SJ (2006). Multisensory activation of the intraparietal area when classifying grating orientation: a functional magnetic resonance imaging study. *J Neurosci*, 26:7491–7501.

Kosslyn SM, Thompson WL, and Alpert NM (1997). Neural systems shared by visual imagery and visual perception: a positron emission tomography study. *Neuroimage*, 6:320–334.

Krubitzer L, Clarey J, Tweedale R, Elston G, and Calford M (1995). A redefinition of somatosensory areas in the lateral sulcus of macaque monkeys. *J Neurosci*, 15:3821–3839.

Krubitzer LA and Kaas JH (1990). The organization and connections of somatosensory cortex in marmosets. *J Neurosci*, 10:952–974.

Lacey S, Flueckiger P, Stilla R, Lava M, and Sathian K (2010). Object familiarity modulates the relationship between visual object imagery and haptic shape perception. *Neuroimage*, 49:1977–1990.

Lacey S and Sathian K (2011). Multisensory object representation: insights from studies of vision and touch. *Prog Brain Res*, 191:165–176.

Lewis JW and Van Essen DC (2000). Corticocortical connections of visual, sensorimotor, and multimodal processing areas in the parietal lobe of the macaque monkey. *J Comp Neurol*, 428:112–137.

Li S, Ostwald D, Giese M, and Kourtzi Z (2007). Flexible coding for categorical decisions in the human brain. *J Neurosci*, 27:12321–12330.

Logan GD and Cowan WB (1984). On the ability to inhibit thought and action: a theory of an act of control. *Psychol Rev*, 91:295–327.

Logothetis NK and Wandell BA (2004). Interpreting the BOLD signal. *Annu Rev Physiol*, 66:735–769.

Lucan JN, Foxe JJ, Gomez-Ramirez M, Sathian K, and Molholm S (2010). Tactile shape discrimination recruits human lateral occipital complex during early perceptual processing. *Hum Brain Mapp*, 31:1813–1821.

Macey PM, Macey KE, Kumar R, and Harper RM (2004). A method for removal of global effects from fMRI time series. *Neuroimage*, 22:360–366.

Matteau I, Kupers R, Ricciardi E, Pietrini P, and Ptito M (2010). Beyond visual, aural and haptic movement perception: hMT+ is activated by electrotactile motion stimulation of the tongue in sighted and in congenitally blind individuals. *Brain Res Bull*, 82:264–270.

Mechelli A, Price CJ, Friston KJ, and Ishai A (2004). Where bottom-up meets top-down: neuronal interactions during perception and imagery. *Cereb Cortex*, 14:1256–1265.

Merzenich MM, Kaas JH, Sur M, and Lin CS (1978). Double representation of the body surface within cytoarchitectonic areas 3b and 1 in 'SI' in the owl monkey (*Aotus trivirgatus*). *J Comp Neurol*, 181:41–73.

Mima T, Nagamine T, Nakamura K, and Shibasaki H (1998). Attention modulates both primary and second somatosensory cortical activities in humans: a magnetoencephalographic study. *J Neurophysiol*, 80:2215–2221.

Miquée A, Xerri C, Rainville C, Anton JL, Nazarian B, Roth M, and Zennou-Azogui Y (2008). Neuronal substrates of haptic shape encoding and matching: a functional magnetic resonance imaging study. *Neuroscience*, 152:29–39.

Mishkin M (1979). Analogous neural models for tactual and visual learning. *Neuropsychologia*, 17:139–151.

Mishkin M and Ungerleider LG (1982). Contribution of striate inputs to the visuospatial functions of parieto-preoccipital cortex in monkeys. *Behav Brain Res*, 6:57–77.

Mountcastle VB (2005). *The Sensory Hand: Neural Mechanisms of Somatic Sensation*. Cambridge, MA: Harvard University Press. 616 p.

Nakamura J, Endo K, Sumida T, and Hasegawa T (1998). Bilateral tactile agnosia: a case report. *Cortex*, 34:375–388.

Nelson RJ, editor (2002). *The Somatosensory System: Deciphering the Brain's Own Body Image*. Boca Raton, FL: CRC Press. 424 p.

Nelson RJ, Sur M, Felleman DJ, and Kaas JH (1980). Representations of the body surface in postcentral parietal cortex of *Macaca fascicularis*. *J Comp Neurol*, 192:611–643.

Ogawa S, Lee TM, Kay AR, and Tank DW (1990). Brain magnetic resonance imaging with contrast dependent on blood oxygenation. *Proc Natl Acad Sci U S A*, 87:9868–9872.

Pandya DN and Seltzer B (1982). Intrinsic connections and architectonics of posterior parietal cortex in the rhesus monkey. *J Comp Neurol*, 204:196–210.

Pei YC, Hsiao SS, Craig JC, and Bensmaia SJ (2010). Shape invariant coding of motion direction in somatosensory cortex. *PLoS Biol*, 8:e1000305.

Penfield W and Boldrey E (1937). Somatic motor and sensory representation in the cerebral cortex of man as studied by electrical stimulation. *Brain*, 60:389–443.

Penny WD, Stephan KE, Mechelli A, and Friston KJ (2004). Comparing dynamic causal models. *Neuroimage*, 22:1157–1172.

Phillips JR, Johansson RS, and Johnson KO (1990). Representation of braille characters in human nerve fibres. *Exp Brain Res*, 81:589–592.

Phillips JR, Johnson KO, and Hsiao SS (1988). Spatial pattern representation and transformation in monkey somatosensory cortex. *Proc Natl Acad Sci U S A*, 85:1317–1321.

Pietrini P, Furey ML, Ricciardi E, Gobbini MI, Wu WH, Cohen L, Guazzelli M, and Haxby JV (2004). Beyond sensory images: object-based representation in the human ventral pathway. *Proc Natl Acad Sci U S A*, 101:5658–5663.

Pleger B, Ruff CC, Blankenburg F, Bestmann S, Wiech K, Stephan KE, Capilla A, Friston KJ, and Dolan RJ (2006). Neural coding of tactile decisions in the human prefrontal cortex. *J Neurosci*, 26:12596–12601.

Posner MI (1986). *Chronometric Explorations of Mind*. New York: Oxford University Press. 271 p.

Powell TP and Mountcastle VB (1959). Some aspects of the functional organization of the cortex of the postcentral gyrus of the monkey: a correlation of findings obtained in a single unit analysis with cytoarchitecture. *Bull Johns Hopkins Hosp*, 105:133–162.

Price CJ and Friston KJ (1997). Cognitive conjunction: a new approach to brain activation experiments. *Neuroimage*, 5:261–270.

Pruett JR Jr, Sinclair RJ, and Burton H (2000). Response patterns in second somatosensory cortex (SII) of awake monkeys to passively applied tactile gratings. *J Neurophysiol*, 84:780–797.

Rao RP and Ballard DH (1999). Predictive coding in the visual cortex: a functional interpretation of some extra-classical receptive-field effects. *Nat Neurosci*, 2:79–87.

Reed CL and Caselli RJ (1994). The nature of tactile agnosia: a case study. *Neuropsychologia*, 32:527–539.

Reed CL, Caselli RJ, and Farah MJ (1996). Tactile agnosia. Underlying impairment and implications for normal tactile object recognition. *Brain*, 119:875–888.

Reed CL, Klatzky RL, and Halgren E (2005). What vs. where in touch: an fMRI study. *Neuroimage*, 25:718–726.

Bibliography

Ricciardi E, Vanello N, Sani L, Gentili C, Scilingo EP, Landini L, Guazzelli M, Bicchi A, Haxby JV, and Pietrini P (2007). The effect of visual experience on the development of functional architecture in hMT+. *Cereb Cortex*, 17:2933–2939.

Rubia K, Smith AB, Brammer MJ, and Taylor E (2003). Right inferior prefrontal cortex mediates response inhibition while mesial prefrontal cortex is responsible for error detection. *Neuroimage*, 20:351–358.

Ruiz S, Crespo P, and Romo R (1995). Representation of moving tactile stimuli in the somatic sensory cortex of awake monkeys. *J Neurophysiol*, 73:525–537.

Sani L, Ricciardi E, Gentili C, Vanello N, Haxby JV, and Pietrini P (2010). Effects of visual experience on the human MT+ functional connectivity networks: an fMRI study of motion perception in sighted and congenitally blind individuals. *Front Syst Neurosci*, 4:159.

Scheperjans F, Palomero-Gallagher N, Grefkes C, Schleicher A, and Zilles K (2005). Transmitter receptors reveal segregation of cortical areas in the human superior parietal cortex: relations to visual and somatosensory regions. *Neuroimage*, 28:362–379.

Schneider RJ, Friedman DP, and Mishkin M (1993). A modality-specific somatosensory area within the insula of the rhesus monkey. *Brain Res*, 621:116–120.

Shikata E, Hamzei F, Glauche V, Knab R, Dettmers C, Weiller C, and Büchel C (2001). Surface orientation discrimination activates caudal and anterior intraparietal sulcus in humans: an event-related fMRI study. *J Neurophysiol*, 85:1309–1314.

Siesjo BK (1978). *Brain Energy Metabolism*. New York: John Wiley & Sons Ltd. 620 p.

Sinclair RJ and Burton H (1991). Neuronal activity in the primary somatosensory cortex in monkeys (*Macaca mulatta*) during active touch of textured surface gratings: responses to groove width, applied force, and velocity of motion. *J Neurophysiol*, 66:153–169.

Sinclair RJ, Kuo JJ, and Burton H (2000). Effects on discrimination performance of selective attention to tactile features. *Somatosens Mot Res*, 17:145–157.

Spitzer B, Wacker E, and Blankenburg F (2010). Oscillatory correlates of vibrotactile frequency processing in human working memory. *J Neurosci*, 30:4496–4502.

Staines WR, Graham SJ, Black SE, and McIlroy WE (2002). Task-relevant modulation of contralateral and ipsilateral primary somatosensory cortex and the role of a prefrontal-cortical sensory gating system. *Neuroimage*, 15:190–199.

Stephan KE and Friston KJ (2010). Analyzing effective connectivity with fMRI. *Wiley Interdiscip Rev Cogn Sci*, 1:446–459.

Stephan KE, Penny WD, Daunizeau J, Moran RJ, and Friston KJ (2009). Bayesian model selection for group studies. *Neuroimage*, 46:1004–1017.

Stephan KE, Penny WD, Moran RJ, den Ouden HE, Daunizeau J, and Friston KJ (2010). Ten simple rules for dynamic causal modeling. *Neuroimage*, 49:3099–3109.

Stilla R and Sathian K (2008). Selective visuo-haptic processing of shape and texture. *Hum Brain Mapp*, 29:1123–1138.

Stoeckel MC, Weder B, Binkofski F, Buccino G, Shah NJ, and Seitz RJ (2003). A fronto-parietal circuit for tactile object discrimination: an event-related fMRI study. *Neuroimage*, 19:1103–1114.

Stoesz MR, Zhang M, Weisser VD, Prather SC, Mao H, and Sathian K (2003). Neural networks active during tactile form perception: common and differential activity during macrospatial and microspatial tasks. *Int J Psychophysiol*, 50:41–49.

Summerfield C and Egner T (2009). Expectation (and attention) in visual cognition. *Trends Cogn Sci*, 13:403–409.

Summerfield C, Egner T, Greene M, Koechlin E, Mangels J, and Hirsch J (2006). Predictive codes for forthcoming perception in the frontal cortex. *Science*, 314:1311–1314.

Summers IR, Francis ST, Bowtell RW, McGlone FP, and Clemence M (2009). A functional-magnetic-resonance-imaging investigation of cortical activation from moving vibrotactile stimuli on the fingertip. *J Acoust Soc Am*, 125:1033–1039.

Talairach J and Tournoux P (1988). *Co-planar Stereotaxic Atlas of the Human Brain: 3-Dimensional Proportional System - an Approach to Cerebral Imaging*. New York: Thieme Medical Publishers. 122 p.

Thompson R and Duncan J (2009). Attentional modulation of stimulus representation in human fronto-parietal cortex. *Neuroimage*, 48:436–448.

Tootell RB, Reppas JB, Kwong KK, Malach R, Born RT, Brady TJ, Rosen BR, and Belliveau JW (1995). Functional analysis of human MT and related visual cortical areas using magnetic resonance imaging. *J Neurosci*, 15:3215–3230.

Van Boven RW, Ingeholm JE, Beauchamp MS, Bikle PC, and Ungerleider LG (2005). Tactile form and location processing in the human brain. *Proc Natl Acad Sci U S A*, 102:12601–12605.

van der Zwaag W, Marques JP, Lei H, Just N, Kober T, and Gruetter R (2009). Minimization of Nyquist ghosting for echo-planar imaging at ultra-high fields based on a "negative readout gradient" strategy. *J Magn Reson Imaging*, 30:1171–1178.

Vega-Bermudez F and Hsiao SS (2002). Attention in the somatosensory system. In *The Somatosensory System: Deciphering the Brain's Own Body Image*. Boca Raton, FL: CRC Press.

Vega-Bermudez F and Johnson KO (1999). Surround suppression in the responses of primate SA1 and RA mechanoreceptive afferents mapped with a probe array. *J Neurophysiol*, 81:2711–2719.

Warren S, Hamalainen HA, and Gardner EP (1986). Objective classification of motion- and direction-sensitive neurons in primary somatosensory cortex of awake monkeys. *J Neurophysiol*, 56:598–622.

Watson JD, Myers R, Frackowiak RS, Hajnal JV, Woods RP, Mazziotta JC, Shipp S, and Zeki S (1993). Area V5 of the human brain: evidence from a combined study using positron emission tomography and magnetic resonance imaging. *Cereb Cortex*, 3:79–94.

Woolsey CN (1943). "Second" somatic receiving areas in the cerebral cortex of cat, dog and monkey. *Fed Proc*, 2:55.

Worsley KJ, Marrett S, Neelin P, Vandal AC, Friston KJ, and Evans AC (1996). A unified statistical approach for determining significant signals in images of cerebral activation. *Hum Brain Mapp*, 4:58–73.

Zeki S, Watson JD, Lueck CJ, Friston KJ, Kennard C, and Frackowiak RS (1991). A direct demonstration of functional specialization in human visual cortex. *J Neurosci*, 11:641–649.

List of Figures

2.1. Cortical regions involved in somatosensory processing 14
2.2. Model of somatosensory processing for perception and action 19

3.1. Mixed-effects analysis for population inference 27
3.2. Conceptual basis of DCM 30

4.1. Illustration of tactile stimuli used 39
4.2. Overall neuronal network associated with tactile stimulation 44
4.3. Stimulus-specific differences and differential effects for motion direction and pattern orientation 46
4.4. Areas involved in feature processing and performance-dependent covariation 48
4.5. PPI analyses for motion and pattern processing 50

5.1. Stimuli and experimental design 59
5.2. Overall neuronal networks involved in the different tasks 63
5.3. Differential effects between the tasks 64
5.4. Differential effects of change in the attended stimulus attribute 66
5.5. PPI analysis for selective attention 68
5.6. Analysis of effective connectivity between SI, SII, and IFG 70

6.1. Extended model of somatosensory processing for perception and action . . 79

A.1. BOLD time courses extracted from SI and SII 88
A.2. BOLD time courses extracted from SI and aSPC 88
A.3. BOLD time courses extracted from hMT+/V5 and IPC 88

B.1. Overlap with anatomical map for hMT+/V5 activation 89
B.2. Functional regions active during motor responses 90

C.1. Bootstrapping correlations 93

List of Figures

List of Tables

2.1. Mechanoreceptive afferents and their properties 12
4.1. Functional regions active during tactile stimulation 45
5.1. Functional regions active during the different tasks 65
5.2. Functional regions active during change detection 67
5.3. Model configurations 71

List of Tables

List of Abbreviations

AIP	anterior intraparietal area
aIPS	anterior intraparietal sulcus
ANOVA	analysis of variance
aSPC	anterior superior parietal cortex
BMS	Bayesian model selection
BOLD	blood-oxygen level-dependent
BPA	Bayesian parameter averaging
DCM	dynamic causal modeling
EPI	echo planar imaging
FFX	fixed-effects analysis
fMRI	functional magnetic resonance imaging
F task	frequency attention task
FWHM	full-width at half-maximum
GLM	general linear model
GNG	go/no-go task
GRF	Gaussian random field
hIP1–3	human intraparietal area 1–3
hMT+/V5	motion-sensitive area in human middle temporal cortex
HRF	hemodynamic response function
IFG	inferior frontal gyrus
IPC	inferior parietal cortex
IPS	intraparietal sulcus
ISI	inter-stimulus interval

List of Abbreviations

LMGS	linear model of global signal
LOC	lateral occipital complex
LPFC	lateral prefrontal cortex
MI	primary motor cortex
MNI	Montreal Neurological Institute
MR	magnetic resonance
MST	medial superior temporal area
MT	middle temporal area
OP 1–4	*operculum parietale* 1–4
PPC	posterior parietal cortex
PPI	psychophysiological interaction
pre-SMA	pre-supplementary motor area
P task	pattern attention task
PV	parietal ventral area
RAI	rapidly adapting type I mechanoreceptive afferents
RAII	rapidly adapting type II mechanoreceptive afferents
RFX	random-effects analysis
RMS	root mean square
S2	second somatosensory area
SAI	slowly adapting type I mechanoreceptive afferents
SAII	slowly adapting type II mechanoreceptive afferents
SEM	standard error of the mean
SI	primary somatosensory cortex
SII	secondary somatosensory cortex
SMA	supplementary motor area
SPC	superior parietal cortex
SPM	statistical parametric map
STS	stop signal task
TE	echo time
TR	repetition time

VIP	ventral intraparietal area
VP	ventroposterior nucleus of the thalamus
VS	ventral somatosensory area

List of Abbreviations

Die VDM Verlagsservicegesellschaft sucht für wissenschaftliche Verlage abgeschlossene und herausragende

Dissertationen, Habilitationen, Diplomarbeiten, Master Theses, Magisterarbeiten usw.

für die kostenlose Publikation als Fachbuch.

Sie verfügen über eine Arbeit, die hohen inhaltlichen und formalen Ansprüchen genügt, und haben Interesse an einer honorarvergüteten Publikation?

Dann senden Sie bitte erste Informationen über sich und Ihre Arbeit per Email an *info@vdm-vsg.de*.

Sie erhalten kurzfristig unser Feedback!

VDM Verlagsservicegesellschaft mbH
Dudweiler Landstr. 99   Telefon +49 681 3720 174
D - 66123 Saarbrücken   Fax   +49 681 3720 1749
www.vdm-vsg.de

Die VDM Verlagsservicegesellschaft mbH vertritt

Printed by Books on Demand GmbH, Norderstedt / Germany